Macmillan/McGraw-Hill Science

THE ANIMAL KINGDOM

AUTHORS

Mary Atwater
The University of Georgia

Prentice Baptiste
University of Houston

Lucy Daniel
Rutherford County Schools

Jay Hackett
University of Northern Colorado

Richard Moyer
University of Michigan, Dearborn

Carol Takemoto
Los Angeles Unified School District

Nancy Wilson
Sacramento Unified School District

*Canada Geese
on their migration route*

Macmillan/McGraw-Hill
School Publishing Company
New York   Chicago   Columbus

MACMILLAN / McGRAW-HILL

Environmental Education:
Cheryl Charles, Ph.D.
Executive Director
Project Wild
Boulder, CO

Gifted:
Dr. James A. Curry
Associate Professor, Graduate Faculty
College of Education, University of Southern Maine
Gorham, ME

Global Education:
M. Eugene Gilliom
Professor of Social Studies and Global Education
The Ohio State University
Columbus, OH

Life Science:
Wyatt W. Anderson
Professor of Genetics
University of Georgia
Athens, GA

Orin G. Gelderloos
Professor of Biology and Professor of Environmental Studies
University of Michigan—Dearborn
Dearborn, MI

Donald C. Lisowy
Education Specialist
New York, NY

Dr. E.K. Merrill
Assistant Professor
University of Wisconsin Center—Rock County
Madison, WI

Literature:
Dr. Donna E. Norton
Texas A&M University
College Station, TX

Derrick R. Lavoie
Assistant Professor of Science Education
Montana State University
Bozeman, MT

CONSULTANTS

Assessment:
Mary Hamm
Associate Professor
Department of Elementary Education
San Francisco State University
San Francisco, CA

Cognitive Development:
Pat Guild, Ed.D.
Director, Graduate Programs in Education and
Learning Styles Consultant
Antioch University
Seattle, WA

Kathi Hand, M.A.Ed.
Middle School Teacher and Learning Styles Consultant
Assumption School
Seattle, WA

Earth Science:
David G. Futch
Associate Professor of Biology
San Diego State University
San Diego, CA

Dr. Shadia Rifai Habbal
Harvard-Smithsonian Center for Astrophysics
Cambridge, MA

Tom Murphree, Ph.D.
Global Systems Studies
Monterey, CA

Suzanne O'Connell
Assistant Professor
Wesleyan University
Middletown, CT

Sidney E. White
Professor of Geology
The Ohio State University
Columbus, OH

Copyright © 1993 Macmillan/McGraw-Hill School Publishing Company

All rights reserved. No part of this book may be reproduced or transmitted in any form or by any means, electronic or mechanical, including photocopying, recording, or by any information storage and retrieval system, without permission in writing from the publisher.

Macmillan/McGraw-Hill School Division
10 Union Square East
New York, New York 10003
Printed in the United States of America

ISBN 0-02-274276-X / 6

3 4 5 6 7 8 9 RRW 99 98 97 96 95 94 93

Mathematics:
Dr. Richard Lodholz
Parkway School District
St. Louis, MO

Middle School Specialist:
Daniel Rodriguez
Principal
Pomona, CA

Misconceptions:
Dr. Charles W. Anderson
Michigan State University
East Lansing, MI

Dr. Edward L. Smith
Michigan State University
East Lansing, MI

Multicultural:
Bernard L. Charles
Senior Vice President
Quality Education for Minorities Network
Washington, DC

Paul B. Janeczko
Poet
Hebron, MA

James R. Murphy
Math Teacher
La Guardia High School
New York, NY

Clifford E. Trafzer
Professor and Chair, Ethnic Studies
University of California, Riverside
Riverside, CA

Physical Science:
Gretchen M. Gillis
Geologist
Maxus Exploration Company
Dallas, TX

Henry C. McBay
Professor of Chemistry
Morehouse College and Clark Atlanta University
Atlanta, GA

Wendell H. Potter
Associate Professor of Physics
Department of Physics
University of California, Davis
Davis, CA

Claudia K. Viehland
Educational Consultant, Chemist
Sigma Chemical Company
St. Louis, MO

Reading:
Charles Temple, Ph.D.
Associate Professor of Education
Hobart and William Smith Colleges
Geneva, NY

Safety:
Janice Sutkus
Program Manager: Education
National Safety Council
Chicago, IL

Science Technology and Society (STS):
William C. Kyle, Jr.
Director, School Mathematics and Science Center
Purdue University
West Lafayette, IN

Social Studies:
Jean Craven
District Coordinator of Curriculum Development
Albuquerque Public Schools
Albuquerque, NM

Students Acquiring English:
Mario Ruiz
Pomona, CA

STUDENT ACTIVITY TESTERS

Alveria Henderson	Andrew Duffy
Kate McGlumphy	Chris Higgins
Katherine Petzinger	Sean Pruitt
John Wirtz	Joanna Huber
Sarah Wittenbrink	John Petzinger

FIELD TEST TEACHERS

Kathy Bowles
Landmark Middle School
Jacksonville, FL

Myra Dietz
#46 School
Rochester, NY

John Gridley
H.L. Harshman Junior High School #101
Indianapolis, IN

Annette Porter
Schenk Middle School
Madison, WI

Connie Boone
Fletcher Middle School
Jacksonville, FL

Theresa Smith
Bates Middle School
Annapolis, MD

Debbie Stamler
Sennett Middle School
Madison, WI

Margaret Tierney
Sennett Middle School
Madison, WI

Mel Pfeiffer
I.P.S. #94
Indianapolis, IN

CONTRIBUTING WRITER

Flora Foss

ACKNOWLEDGEMENTS

DEAR BRONX ZOO by Joyce Altman & Sue Goldberg (New York: Macmillan, 1990).

THE INCREDIBLE JOURNEY by Shelia Brunford. Copyright © 1960, 1961 by Shelia Burnford. Copyright renewed © 1988 by Jonquil Graves, Juliet Pin and Peronelle Robbins.

Buffalo

THE ANIMAL KINGDOM

4

Activities!

EXPLORE

TRY THIS

Features

Links

Departments

THE ANIMAL

KINGDOM

*T*hroughout history, people have been fascinated by animals. Animal life is found in almost every environment on Earth. Each animal contributes to the environment and has a specific role there. Humans are only one of the many species, or kinds, of animals. By studying animals, you can

discover how you fit into the network of living things that coexist on Earth.

All life is interdependent. How does an animal such as the mosquito have an effect upon you? What are other animals that affect your daily life?

Wherever you are on Earth, there are animals near you. Over a million species of animals exist on Earth today. Many others have become extinct. How did so many different species of animals come to be? Why did others disappear? People have often wondered about all of the various species of animals and have developed methods to study animals of today and compare them with animals that existed throughout history.

The animal kingdom consists of living things ranging from mosquitos and corals to cougars.

Minds On! Examine the different animals pictured here. Do they resemble any animals you've seen? These animals lived in prehistoric times and no longer exist on Earth. Why, then, can you see similarities between prehistoric animals and animals of today? Try to list a present-day animal for each of these prehistoric animals on page 1 in your *Activity Log.* What are the similarities and differences between the prehistoric and present-day animals? Why do you think these animals are similar? ●

How do you think prehistoric animals resemble those of today?

Because Earth has changed over time, environments on Earth have also changed. Animals have adapted to new living conditions. These adaptations have occurred very gradually over millions of years.

Scientists know that prehistoric animals existed by studying fossils. Because scientists can determine when a particular fossil formed, they know when various animals existed on Earth.

Minds On! Today, some animals are kept in zoos. There, scientists study their habits, behaviors, and life cycles. Have you ever wondered why certain animals are placed in zoos? If you were to create a zoo, what animals would you have there? Be sure to consider the city where the zoo would be located, the housing, food, and other elements you'd have to supply that would affect the animals. Write your ideas for creating a zoo on page 2 in your *Activity Log*. Discuss your ideas with others and compare your dream zoos. ●

Now let's investigate all the kinds of animals alive today. How can we begin to do this? First, we must know how scientists group animals. Scientists classify living things into groups based on shared evolutionary characteristics. The largest group of living things is called a kingdom. Each kingdom is divided into smaller groups. The largest group within a kingdom is a phylum. The animal kingdom is classified into many phyla, nine of which will be studied in this unit. Phyla are divided into classes. Classes are further divided into smaller and smaller groups. The smallest group is the species. Members of a species are able to reproduce among themselves.

As we investigate animal phyla, we'll have many questions. Sometimes the answers can be found in books. Turn the page to see some of the books that answer questions about animals.

SCIENCE IN
LITERATURE

The great variety of animal life demonstrates very well the power of evolution. How have so many kinds of animals come to be? Their environments shaped them over time, until now almost every individual microclimate and microecosystem contains animal life that is perfectly adapted to it. How many ways are there for animals to catch food, to survive in all kinds of weather, to reproduce, and to raise young? There are probably as many ways as there are animal species. Learn about animals and their similarities and differences in the following books.

Dear Bronx Zoo **by Joyce Altman and Sue Goldberg. New York: Macmillan, 1990.**

"What happens when a giraffe gets a sore throat?" "How big are baby gorillas when they're born?" "Does a sea cow give milk? "How can I help protect wildlife?" These and many other questions are answered by the Friends of the Zoo. Learn how animals live in nature and in zoos, how the zoo can care for over 4,200 animals' needs, and how zoos today can protect endangered species.

The Incredible Journey by Sheila Burnford. New York: Bantam Books, 1961.

This heartwarming novel traces the journey of three pets across northern Canada in late autumn. An elderly bulldog, a young Labrador retriever, and a Siamese cat travel together and make their way in spite of hardships that include wild animals, unfriendly humans, and harsh weather. What humans and animals help the three travelers on their way? Do they succeed in regaining their lost human family? Only *The Incredible Journey* can tell you for sure.

OTHER GOOD BOOKS TO READ

101 Questions and Answers About Dangerous Animals by Seymour Simon. New York: Macmillan, 1985.

How big are the dangerous cats? How dangerous are wolves? Are vultures dangerous? How large is a colony of army ants? What is the most dangerous animal in the world? Ask Seymour Simon and get clear, complete, and entertaining replies. Illustrated with black and white line drawings.

What Do Animals See, Hear, Smell, and Feel? by Ranger Rick Books. National Wildlife Federation, 1990.

How does the octopus see from inside its cloud of inky dye? Do barnacles and shrimps have ears? Do bees and birds really see a color humans can't see? How do fish make noises to herd themselves into schools? Answers to these and hundreds more questions about animal senses are to be found within the covers of this reference book.

Animal Camouflage: A Closer Look by Joyce Powzyk. New York: Bradbury Press, 1990.

How are animals adapted for camouflage? Some resemble natural objects like twigs. Some look like and behave like other animals. Some have patterns that break up their outlines and make their shapes hard to see. Some blend in with their surroundings. And then there are the really unusual adaptations! Read about them all in this beautifully illustrated book.

INVERTEBRATES

What makes an animal an invertebrate? You can begin to find the answer by observing animals.

Starfish are one kind of invertebrate.

In order to study organisms, scientists first observe their characteristics. They group organisms based on the kind of characteristics they share. For instance, all organisms in the animal kingdom are made of more than one cell and must obtain their own food. Their cells lack a cell wall and have a nucleus. Because these characteristics are shared by all animals, scientists infer that all animals evolved from a common ancestor.

After an organism has been classified as an animal, scientists look for other characteristics in order to place it in a phylum. All animals with backbones are grouped in the same phylum. You will study these animals in the next lesson. In this lesson you will look at phyla of animals that lack a backbone. Together, these animals are known as invertebrates. Each separate phylum is made up of animals that share characteristics derived from a common ancestor.

Ladybugs are one kind of animal that can fly.

Roundworms are
often found in the
soil or in other
organisms

Clams live in a water
environment.

13

Activity!

Invertebrates: Sorting Out the Differences

In this Explore Activity, you'll study two different invertebrates and compare their characteristics.

What You Need

earthworm
2 planarians
small paintbrush
small piece of liver
jar lid
water
2 moist paper towels
moist soil and sand
hand lens
aluminum pan
Activity Log pages 3-4

What To Do

1 Fill the jar lid half full of water.

2 Use the paintbrush to transfer 2 planarians from the culture to the jar lid. *Safety Tip:* Handle live animals gently and carefully.

earthworm

planarian

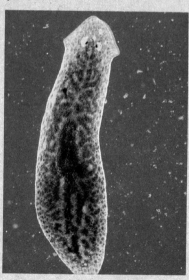

Safety!

See the *Safety Tips* in steps 2 and 8.

3 Observe the planarians with the hand lens, and record your answers to the questions in the table in your *Activity Log*.

4 Place a piece of liver about the size of a pinhead in the jar lid, and observe how the planarian feeds. Answer the questions in the data table.

5 Draw how the planarian looks in your *Activity Log*.

6 Return the planarian to the place your teacher has chosen.

7 Line your pan with a moist paper towel, and place a cup of the moist soil into one end of the pan.

8 Wet your hands and place an earthworm on the moist paper towel in the pan.

Handle live animals gently and carefully. Be careful not to let the earthworm dry out. Keep your hands moist while touching it.

9 Use the hand lens to observe the earthworm and answer the questions in the table in your *Activity Log*.

10 Place the earthworm on the soil. Touch the earthworm gently. Record your observations in your *Activity Log*.

11 Return the earthworm to the place your teacher has chosen and wash your hands with soap and water.

DATA TABLE

	Planarian	Earthworm
Can you find the head and tail?		
Are there any sense organs on the head?		
How does it move?		
Where is the mouth?		
Does it have one or two openings in its digestive system?		
Do you see any evidence of blood vessels or observe a pulse?		
What do you think is the most likely habitat for this animal?		

What Happened?

1. From your observations, explain how the earthworm's body is different from the planarian's.
2. From what you could observe, how does the earthworm's method of feeding differ from the planarian's?
3. How are the earthworm and planarian similar?

What Now?

1. You noted differences and similarities in the structure and behavior of two animals. These animals are classified into different phyla. Why do you think this is so?

EXPLORE

15

FOUR PHYLA OF
INVERTEBRATES

In the Explore Activity, you looked at two phyla of invertebrates. You saw they had some characteristics in common, but many differences. In this part of the lesson, you will look more closely at flatworms, the phylum planarians belong to, as well as three other phyla—sponges, cnidarians, and roundworms. As you study each group, think about the similarities and differences each shares with the others.

Sponges have no organ systems and no true tissues. Many have no definite shape and live attached to one spot. Sponges have two layers of cells with a jelly-like substance between the layers. The cells are specialized to perform different tasks. You can examine a dried natural sponge in the next Try This Activity.

Sponges have many holes or pores that carry water throughout their bodies. Water flows in through the pores and out at the top of the sponge.

Most sponges are found in salt water and vary in size from one centimeter (about half an inch) in height to over two meters (about six feet).

Activity!

Artificial or Natural?

Compare an artificial kitchen sponge with a sponge that was once a living animal.

What You Need

artificial sponge, natural sponge, hand lens, *Activity Log* page 5

Examine each sponge. How do they feel? Observe both sponges with a hand lens. Record your observations in your *Activity Log.* How are the natural and artificial sponges alike? How are they different? Which kind do you think would be less expensive? Why?

Fossils of sponges that existed 570 million years ago have been found. Sponges were the first animals to evolve a many-celled body with each cell having a specific function.

Cnidarians (nī dâr′ ē əns) are hollow-bodied organisms with stinging cells. Like sponges, cnidarians have two layers of cells. Unlike sponges, the cells are organized into tissues. Tissues are groups of similar cells that work together to perform a specific function.

Stinging cells are located on tentacles near the mouth. The stinging cells help the organism by paralyzing its prey. The tentacles bring the prey to the mouth.

One type of cnidarian is the jellyfish. Other members include corals and sea anemones. Have you ever seen a jellyfish? Its body is largely fluid; 96 percent of its weight is water. Could this animal live out of the water?

Cnidarians were the first organisms to evolve tissues, a definite shape, muscle and nerve fibers, and a mouth that leads to a digestive cavity.

17

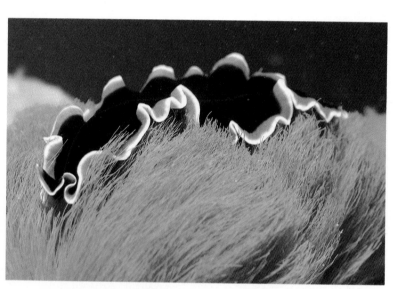

Flatworms have organ systems. They also have their senses and nerves concentrated in the head region.

Flatworms have a head, a tail, and a long flat body. In addition to tissues, flatworms have organs and organ systems, including a digestive system with one opening, reproductive systems, and a nervous system. They live in water or inside the body of another organism. Animals that live inside another organism and obtain food from that organism are parasites.

Two kinds of flatworms are planarians and tapeworms. You looked at the characteristics of planarians in the Explore Activity. They live in water. Tapeworms are flatworms that are parasites. Some tapeworms can live in the digestive tract of humans.

Roundworms have a tube-like body tapering to a point at each end. They also have organ systems, including a digestive system with two openings, a nervous system, and reproductive systems. They are found in soil, in animals, in plants, and in fresh and salt water. Some roundworms, such as hookworms and trichina (tri kī′ nə) worms, are parasites. These roundworms can affect humans. Hookworms enter the body through the skin, and humans can become infected with trichina worms by eating undercooked pork. Most roundworms, however, are free-living and don't depend on other organisms for food or a place to live.

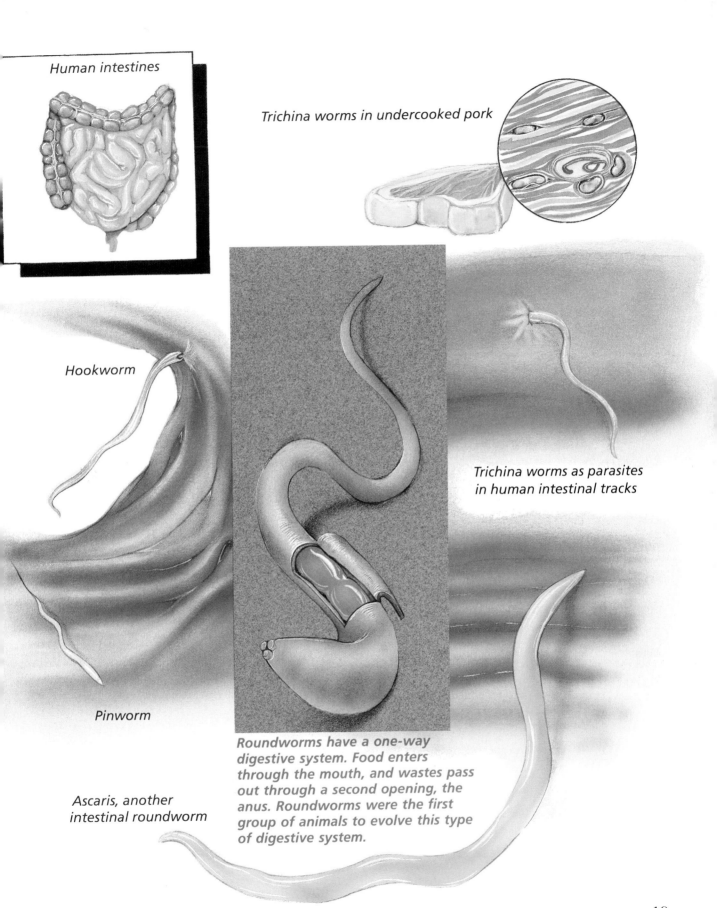

Human intestines

Trichina worms in undercooked pork

Hookworm

Trichina worms as parasites
in human intestinal tracks

Pinworm

Roundworms have a one-way
digestive system. Food enters
through the mouth, and wastes pass
out through a second opening, the
anus. Roundworms were the first
group of animals to evolve this type
of digestive system.

Ascaris, another
intestinal roundworm

INVERTEBRATES

Y ou have now studied four phyla of invertebrates. In this part of the lesson, you'll be examining four more phyla. These animals all have organ systems. The systems in these animals have more structures than those you have just studied. Study of similar organ systems in different animals helps scientists determine evolutionary relationships and gives scientists a basis for grouping animals.

Mollusks are soft-bodied animals that live on land or in fresh or salt water. Most mollusks have a head with eyes and sensory cells that receive stimuli from their environment.

Because mollusks have soft bodies, they need protection. What do you think mollusks have to protect their soft bodies? Most have shells that act as shields and prevent their bodies from drying out. The shell is formed by a fold of tissue called the **mantle** that surrounds the mollusk's soft body.

Mollusks move in order to obtain food, which they take in through the mouth. Some mollusks also have a radula, a tongue-like organ with rows of teeth, used for scraping and tearing food. Mollusks are categorized based on the kind of foot they have and whether or not they have a shell. For example, slugs have no shell, snails have one shell, clams and oysters have two shells with a hinge, and squids have an internal shell.

To observe mollusks' traits for yourself, do the activity on the following page.

Have you ever seen mollusks move? They do so with the use of a muscular foot. This foot releases a trail of slime upon which the animal moves.

Activity!

What Mollusk Traits Does a Squid Have?

Now you can observe a squid to see its mollusk traits.

What You Need

squid
metal tray
forceps
hand lens
Activity Log page 6

Place the squid into the tray with the top side up. See if you can find the body, head, eyes, tentacles, mouth, fins, and mantle. Draw and label the parts of the squid in your *Activity Log.* Use the hand lens to examine the suction cups on the tentacles. What might be the function of these cups? Next, locate the dark jaws in the center of the mouth. Use the forceps carefully to remove the top part of the jaw and then the bottom part. Place the two parts together to observe how they move.

A squid swims by squirting water from its water jet. Locate the water jet. Next, find the hard point at the end of the squid's fin. This is the internal shell, called the pen. Grasp it with the forceps and pull it straight out from the squid's body. What mollusk traits have you observed from examining this squid? List these traits in your *Activity Log* under your drawing.

Scientists have concluded that mollusks probably share a common ancestor with segmented worms. The embryos of both mollusks and segmented worms are similar. **Segmented worms** are animals with bodies that are divided into rings or segments and a coelom.

You observed segmented worms when you examined the earthworm in the Explore Activity on pages 14 and 15. Segmented worms have a specialized digestive system and a closed circulatory system. When a circulatory system is closed, the blood is contained in vessels and is transported throughout the body.

Coelomic cavity

A coelom (sē′ ləm) is a fluid-filled space in which organs develop. Scientists hypothesize that segmented worms were the first organisms to evolve a coelom.

Arthropods make up the next group you'll study. There are over 850,000 kinds of arthropods on Earth. An **arthropod** is an animal with jointed appendages or legs, a segmented body, and an outer skeleton called an exoskeleton. The exoskeleton protects the body and prevents it from drying out. What other invertebrate have you studied that has this characteristic? The exoskeleton of an arthropod is made of nonliving material, so it doesn't grow as the animal grows. What must happen to the exoskeleton in order for the animal to grow and develop? The exoskeleton is shed from time to time and replaced by a new one in a process called molting. After molting has occurred, the new exoskeleton is soft and takes a while to harden. During this time the animal isn't well protected. Molting occurs several times during a lifetime.

Arthropods were the first group of animals to live on land and also the first to have jointed appendages such as legs, claws, and antennae.

Antennae are sensory appendages found on the head.

Minds On! Why do you think arthropods were the first animals to live on land? What advantages do they have over other invertebrates to enable them to survive out of water? Record your answers on *Activity Log* page 7. ●

There are several classes of arthropods. How

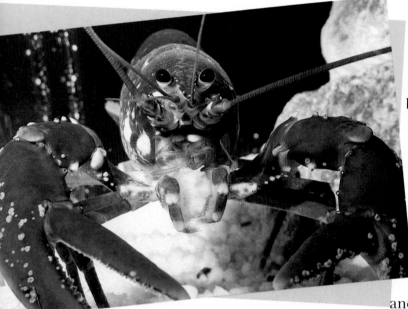

European blue lobster

Insects evolved antennae as sensory organs.

What are some of the ways crabs use their claws?

many are familiar to you? Centipedes and millipedes are two that look like segmented worms with legs. Crabs, crayfish, shrimp, and lobsters belong to another class of arthropods known as crustaceans (kru stā′ shəns). Insects form the largest class of arthropods and live almost everywhere on Earth. Arachnids (ə rak′ nidz) make up another class of arthropods which includes spiders, scorpions, ticks, and mites.

Each of these classes of arthropods has undergone adaptations to suit them to their environment. What is different about each group? What is similar?

Millipedes can have more than 115 pairs of legs. This millipede is found in Equador.

The tarantula is an arthropod belonging to the arachnid class. It is a large, hairy spider. The hairs on a tarantula are part of its sensory system. Even with eight eyes, it can't see well. The hairs are used to feel vibrations and help the tarantula sense its environment. Some hairs allow the spider to taste, and others are used as a defense.

*The tarantula has a well-developed nervous system with a brain, hairs for sensing, and a nerve network that connects to the eyes. The tarantula even breathes with lungs, although they are quite different from your lungs. They're called **book lungs** and consist of many thin leaves of tissue that hang in a blood-filled cavity. The name book lungs comes from the fact that they resemble pages of books.*

The digestive system leads from the mouth to the stomach and gut. The tarantula seizes its prey with fangs and injects venom into the prey to first paralyze and then dissolve it. The liquid material is then sucked from the prey and digested by the tarantula.

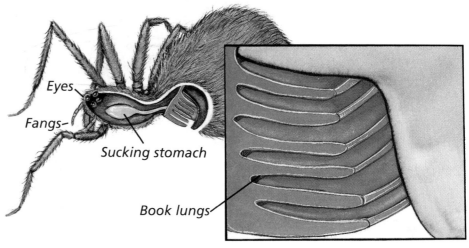

Eyes

Fangs

Sucking stomach

Book lungs

The last phylum you'll study in this lesson is **echinoderms** (i kī′ nə dûrms′), the spiny-bodied animals. These animals live only in oceans and have a water vascular system, that is, a network of water-filled vessels radiating from the mouth area. This system moves water through the animal. It is linked to tube feet, similar to suction cups, that help the animal move.

Examples of echinoderms are sea stars, sea urchins, sand dollars, and sea lilies. These animals have an internal skeleton with spines. Fossil evidence along with the evidence of similar body chemicals and similar development of embryos indicates that echinoderms share a common ancestor with vertebrates.

Echinoderms are
known to have
existed as early as 500
million years ago.

INTERACTING WITH INVERTEBRATES

Minds On! Now that you know more about invertebrates, make a list of invertebrates on page 8 in your *Activity Log.* Beside each one, write why it is important to us. Does it help or harm us? What would happen if it no longer existed? •

Many invertebrates are helpful to humans. For example, ladybugs eat aphids, which are insects that can damage crops.

Health **Link**

TRAPPING DEADLY FLIES

African tsetse fly

While some invertebrates are helpful to humans, others can be harmful. What are some animals that can cause harm to you?

One harmful insect is the tsetse (tset´ sē) fly, because it carries a disease called sleeping sickness that kills many people. Organisms that cause this disease live in tsetse flies, and a person who is bitten may become infected. If the organisms are passed on to a person, they continue their life cycle and can cause the death of the infected person.

In some regions of the world, tsetse flies are a major threat. In Africa, an epidemic of sleeping sickness earlier this century killed 200,000 people. One way Africans are fighting back is by creating traps for tsetse flies. One cost-effective design is made from a plastic bag, staples, and cloth. Some of these traps include colorful shapes and odors to attract the flies. Such a trap can capture up to 20,000 flies per week.

Design your own fly trap for houseflies. With two or three other people, brainstorm about the design of your trap. Make a drawing of what it would look like. Then, describe to the class how it would work.

You've learned that invertebrates such as tsetse flies affect people in harmful ways. However, the opposite is also true. People can have a harmful effect upon invertebrates. One example is the destruction of coral reefs.

CORAL REEFS: CAN THEY SURVIVE?

Most people think of coral as a rock. Actually, coral reefs are made up of living animals, cnidarians, which you have studied. Today, many different factors are harming these delicate creatures. In order to keep them from disappearing, steps must be taken to increase our knowledge about coral reefs and the effects people have on them.

The last remaining coral reefs near the continental United States are located along the southern and eastern coastal areas of Florida. Parts of these reefs extend to depths of 100 meters (about 300 feet), but other parts are found in shallow waters, where they come in contact with boaters. Damage from anchors and propellers causes death to many corals. Even divers may get too close and destroy them. The harm people do to corals takes the reefs hundreds of years to repair.

What else can you think of that may affect coral reefs? Corals require bright, clean water. What effect do you suppose heavy sediment loads and oil spills have on corals?

The future of coral reefs doesn't depend on people alone. The weather is a big factor, too. The shock of waves and wind action is absorbed by the corals and often dislodges them from the reef or breaks them apart. Hurricanes along the Florida coast also harm corals.

Radiocarbon dating of various reefs indicates that coral reefs began forming about 7,000 years ago. If we assume there was one hurricane every seven years, then the coral reefs have been subjected to harmful waves and high winds more than 800 times during their existence. Yet, they have survived. In fact, these reefs have helped to protect the coastlines from devastating storms for many centuries.

One naturally-occurring event is the spread of disease among corals. One coral killer is black band disease. Cyanobacteria become so thick that they appear black. This black mass circulates through the water and clings to corals, turning them a chalk-white color. Black band disease can kill a 100-year-old coral in six months. Experiments are now being conducted with clay to see if corals can be prevented from dying. Any wounds that have been inflicted on corals are patched with clay. The results so far have been positive.

People and natural events affect coral reefs. The more scientists learn about them, the more preventive measures can be taken to protect these beautiful animals.

SUM IT UP

Through your observations of the squid, the earthworm, and the planarian, you were able to recognize similarities and differences between these invertebrates. You've developed an awareness of how animals have evolved from simple creatures to more complex ones. These animals may seem quite simple when compared to humans, but their cells function in the same way ours do to maintain life and promote growth.

CRITICAL THINKING

1. The sponge industry has become smaller in the last 50 years. Can you suggest a reason why?

2. What advantages do parasites have over other animals? What are the disadvantages of parasitic animals?

3. Suggest a reason why sea stars that wash onto the beach cannot crawl back into the water.

4. The word *mollusk* comes from a Latin word meaning "soft." Why is it appropriate for animals such as snails, clams, and squid?

5. After molting but before a new exoskeleton is formed, an arthropod takes in extra water or air that causes its body to swell. Can you suggest why it does this?

VERTEBRATES

What is supporting your body? Is this feature found in any other animals? Without a backbone, you might be just as helpless as a jellyfish out of water.

Minds On! What happens when you stroke a cat's back? What do you feel when you run your hand down a dog's back? Perhaps you have watched a fish swim, a bird fly, a frog leap, or a snake wriggle on the ground. What is similar about all of these animals? What is different? Think about each animal's internal framework, skin type, limbs, and environment. List these similarities and differences on page 9 in your *Activity Log.* ●

Imagine that you are a scientist and you need to group animals according to their similarities and differences. How would you start? Would you look at the traits of each animal? What traits set each animal apart from the others?

As you may have determined, cats, dogs, fish, birds, frogs, and snakes are vertebrates—animals with backbones. All animals with backbones belong to one phylum. However, the animals within this phylum are very different from one another. What traits do you think are used to place vertebrates into classes? The animals you've been thinking about are all categorized into classes according to their traits. Let's examine the different classes of vertebrates.

What do all of these animals have in common?

Activity!

What Are the Characteristics of Vertebrates?

By examining five different classes of vertebrates, you can compare and contrast them. Which animal has developed advanced traits? Which animal best adapts to its environment? To answer these questions, explore each of the following classes of vertebrates.

What You Need

Activity Log pages 10-11

What To Do

1 Observe each animal photograph carefully. Note the size of the animals in the photographs by reading the scale provided for each. Make a data table to answer the following questions.

Giraffe

Iguana

60 cm

15 cm

Amazon parrot

Angelfish

3.5 cm

3 cm

1 cm

Salamander

2
- What is the animal?
- What is its length in centimeters?
- What color is the animal?
- Does it have appendages? If so, what kind and how many?
- What kind of outer covering does it have?
- How do you think it moves?
- What type of environment does it need?

What Happened?

1. Based on your observations, what characteristics do all vertebrates have in common?
2. What traits do each of the animals you studied (fish, amphibian, reptile, bird, mammal) have?

What Now?

1. What are the major differences you observed among fish, amphibians, and reptiles?
2. What are the major differences you observed between birds and mammals?

VERTEBRATES:
THE SKELETAL CONNECTION

The main trait uniting the five classes of vertebrates you studied is that they all have an endoskeleton, an internal skeleton with a backbone. This backbone is made up of small bones called **vertebrae,** which fit together to protect the nerve or spinal cord.

The endoskeleton supports and protects the internal organs of the body. It also provides a place for muscles to attach. To understand how vertebrae are constructed, do the activity below.

Another attribute that is useful to understand when studying vertebrates is the way in which an animal's body temperature is regulated. Have you ever observed a frog on a cold day? It remains still or moves very

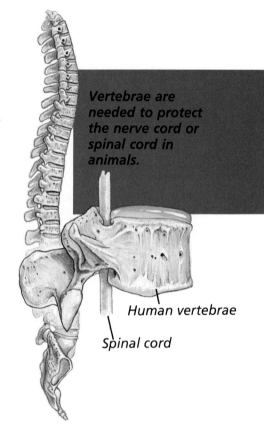

Vertebrae are needed to protect the nerve cord or spinal cord in animals.

Human vertebrae

Spinal cord

TRY THIS Activity!

All I'm Asking for Is a Little Support

Invent your own vertebrate animal and construct its skeleton to help you understand how a skeleton provides support.

What You Need
various materials
Activity Log page 12

Invent an imaginary vertebrate. Draw a picture of the animal and label its parts.

Next, choose the materials you need and lay out the animal's skeleton on its side on newspaper. Build the animal from the bottom up. Build the feet, then attach the legs, and so on.

Think of a name for your animal. Describe where it lives, how it moves, what it eats, and how it would look if it were alive. Choose one member of the group to write a description such as a museum or zoo might display. Make your own exhibit using the animal skeleton, your sketch, and your description.

slowly. How does it feel when you touch it? If it sits on a rock in the sun on a warm day, it becomes warm and can then move quickly. What effect does a colder or warmer temperature have on a gerbil or a dog? In the cold, these animals may shiver or move about more actively. In the heat, they may pant. But their body temperatures remain about the same. This response to temperature is a key difference among vertebrates.

The body of a **cold-blooded animal** changes with the surrounding temperature. A cold-blooded animal relies on heat from its environment to keep the right temperature for life processes. The body temperature of a **warm-blooded** animal stays almost the same, even if the surrounding temperature changes. The animal's body produces heat to maintain an even temperature. Read the Literature Link to learn more about warm-blooded animals.

What are the similarities and differences between this boy, this girl, and their pets?

Literature Link   THE INCREDIBLE JOURNEY

You have read *The Incredible Journey* by Sheila Burnford, in which an old bulldog, a young Labrador retriever, and a Siamese cat find their way home across Canada in autumn. You read that, "Their favorite sleeping places were hollows under uprooted trees where they were sheltered from the wind, and able to burrow down among the drifted leaves for warmth."

How do warm-blooded animals keep warm in winter? If you live where it's never cold, then ask yourself how warm-blooded animals keep cool in summer.

Do research to discover how the local animals maintain their body temperatures despite extremely hot or cold weather. Do large animals have different solutions from those of small animals? Present what you learned to your class. You may give an oral report, publish your findings on the bulletin board, build samples of the animal shelters you found, draw the shelters on a poster, or even write a poem or story that shows how the animals shelter from harsh weather.

FISH: THE WATER DWELLERS

One example of a jawless fish is the lamprey.

Most fish are cold-blooded vertebrates that live in water. Can you think of some things that make fish different from yourself?

How do fish protect themselves? Have you ever felt the skin of a fish? How did it feel? Most fish have scales, which are thin, flat, hard plates that cover the body and help protect it.

Do you know how many kinds of fish there are? Scientists classify fish living today into three classes. Jawless fish have a round mouth, a tubelike body, single fins, no scales, and a skeleton of cartilage. Cartilage is a tough, flexible tissue that's hard, but not as hard as bone.

Cartilage fish have movable jaws, scales, paired fins, and skeletons made of cartilage. Some examples are sharks, skates, and rays.

SCIENTISTS CONCLUDE THAT FISH WERE THE ONLY VERTEBRATES ON EARTH FOR ABOUT 150 MILLION YEARS.

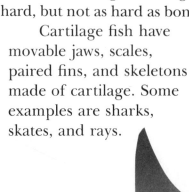

Manta rays live in the open sea and feed on small sea animals.

Bony fish have scales, well-developed sense organs, fins, and skeletons made of bone. Perch, flounder, tuna, and goldfish are bony fish.

Fish first evolved on Earth about 540 million years ago. The first known vertebrates were small, jawless ostracoderms (ôs trak′ ə dermz). From the fossil record, scientists have observed that ostracoderms, like today's jawless fish, lacked teeth.

The cartilage fish and bony fish probably evolved from ancient armored fish called placoderms. They appeared over 425 million years ago.

Fish were the first vertebrates to evolve a two-chambered heart, jaws, teeth, and paired fins. Some bony fish even evolved a lung for breathing. From the fossil record, scientists conclude that fish were probably the only vertebrates on Earth for about 150 million years.

Ostracoderms had large, bony headshields and lacked teeth.

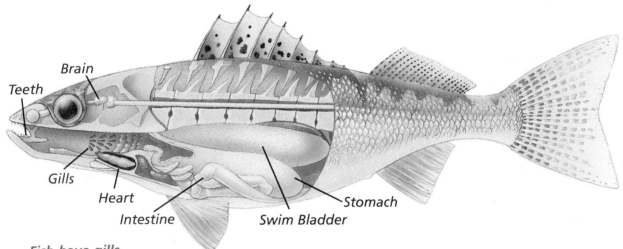

Brain

Teeth

Gills

Heart

Intestine

Stomach

Swim Bladder

Fish have gills, feather-like structures that remove oxygen from the water and release carbon dioxide.

Many bony fish have a swim bladder, which is a thin-walled sac that can be filled with air or deflated to control how deep the fish is in the water.

Fish move through water using fins. Their streamlined bodies allow them to swim through water relatively smoothly. Fins can be single or paired. Fins also help fish balance themselves in the water.

Amphibians: Living in Water and on Land

Amphibians spend part of their lives in water and part on land. Amphibians have moist, smooth skin that allows for the exchange of oxygen and carbon dioxide. They are cold blooded.

When amphibians are on land, how do they keep from drying out? There are mucus glands in an amphibian's skin. Some amphibians have another feature that protects them from other land animals—poison glands.

What amphibians do you know? There are three types of amphibians today —frogs and toads, salamanders, and caecilians (si sil′ yənz). Caecilians are found underground. They have no legs and usually no eyes. Because they exist underground where there is little or no light, caecilians don't need eyes.

Amphibians appeared about 360 million years ago. They were the first vertebrates to have sense organs adapted for living on land. They were the first class of animals to evolve three-chambered hearts, strong muscles, and bones for support and movement.

Over time, amphibians evolved that were less dependent on water. Some returned to the water only to reproduce.

Poison glands in the skin of amphibians produce a bad-tasting substance that keeps predators from attacking them. This Poison Arrow frog from Peru produces poisons that can kill other animals.

Amphibians appeared on Earth 360 million years ago. They're thought to have evolved from lobe-finned fish called crossopterygians (krôs op tə rij′ ēən). These animals had no gills. Instead, they had internal nostrils and a lung that enabled them to stay alive on land for short periods of time.

Reptiles: The Move to Land

Reptiles are cold blooded and have thick, dry, waterproof skin covered with scales. As a reptile grows, its skin is shed by molting.

The four orders of reptiles living today are turtles, alligators and crocodiles, lizards and snakes, and tuataras. Tuataras (tū′ a tä′ raz) are lizard-like reptiles found on small islands near New Zealand.

Reptiles evolved about 310 million years ago. They were able to live on land from birth to death without returning to the water to reproduce. The reproductive process for reptiles begins with internal fertilization. Many reptiles then lay fertilized eggs that have tough, leatherlike coverings. Reptiles were the first vertebrates able to live their entire lives on land.

Reptiles have two pairs of legs with five toes on each foot. Various reptiles use these legs to run, crawl, climb, or paddle.

Tuataras are found on islands near New Zealand.

Minds On! Imagine you are walking in the desert and you hear a rattling buzz. You immediately stop, because you're being warned by a rattlesnake to stay away. What kind of animal is a rattlesnake? How do you think it survives in the desert? What characteristics do other animals have that allow them to survive in the desert? Answer these questions on page 13 in your *Activity Log*. ●

Rattlesnakes are a good illustration of the traits of cold-blooded vertebrates. They gain the heat they need from their surroundings rather than generating heat internally from body processes. However, like all cold-blooded animals, snakes become colder if their environment loses heat. This means their body processes and movements slow down as their temperature decreases. Can you see why many snakes live in deserts? Snakes that live in regions where winter is severe, such as Canada, usually hibernate. Their body processes slow as their temperature drops, and they maintain the lowest possible level of activity.

Rattlesnakes shake their tails to send warnings to enemies.

REPTILES DON'T HAVE TO RETURN TO WATER TO REPRODUCE, AS AMPHIBIANS DO.

A rattlesnake's body shows how well these vertebrates are adapted for life on land. Their tough skin is covered with scales to keep water in. You remember that the skin of an amphibian allows moisture to pass in and out readily. What would happen to an amphibian in a desert environment?

Rattlesnakes get their name from the rattles found on their tails. Like most reptiles, rattlesnakes shed their

skin periodically. When they molt, a button on the tail plus a ring of old skin remain. A new ring is added each time a snake sheds its skin. The attached rings are like stacked cups. When rattlesnakes shake their tails, the rings hit against each other and make a dry, rattling sound.

Rattlesnakes breathe air and have a heart with three chambers. Unlike most reptiles, who produce eggs in watertight, leathery shells, rattlesnakes produce live young. As you know, reptiles don't have to return to water to reproduce as amphibians do. Rattlesnakes lack the four legs and five-clawed feet that most reptiles have. Why, then, are rattlesnakes classed as reptiles, since they lack this feature? Scientists have evidence that the ancestors of rattlesnakes—in fact, of all present-day snakes—did have four legs. Fossils of these ancestors reveal creatures much like today's lizards. Some of these ancestors began to evolve smaller and smaller legs 135 million years ago. Finally, they lost their legs altogether. The pythons and boa snakes of today still have parts of hip bones in their skeletons.

You can see by looking that snakes are legless. Are they earless as well? How do snakes hear as well as they do with no external ears? Do the Try This Activity on page 40 to investigate how snakes sense sounds.

The fangs of the rattlesnake are used to inject poison into prey.

The clawed feet of most reptiles make them well adapted to travel on land. How do land-living snakes travel? Study the diagram to find out.

All snakes are predators. Some snakes squeeze or constrict their prey to strangle it. The rattlesnake is a pit viper, one of a group of snakes that kill their prey by striking it with sharp, hollow fangs and injecting a strong poison.

When snakes devour their prey, they show another helpful adaptation. Snakes have thin ribs and a flexible backbone with 100 to 400 vertebrae.

Snakes contract their muscles in alternate bands. This causes the body to draw up into an S shape. The body wriggles and presses against stones and stems, enabling the snake to push itself forward. Snakes use another kind of muscle contraction and the wide scales on their underside to move straight forward very quietly so they can hunt.

Activity!

Tuning in to Snake Hearing

What You Need
tuning fork
Activity Log page 14

Hold the tuning fork by the stem and tap it on a wooden chair or table. This will cause it to vibrate rapidly. Be careful not to touch the prongs of the tuning fork after you tap it on the table. Hold the tuning fork next to your ear. Describe what you hear. Tap the tuning fork again. Place the base of the stem against your chin. Describe what happens.

In your *Activity Log,* explain how you think a snake hears. Why would this method be more useful for animals living underground than the way humans hear?

This flexibility allows snakes to swallow their prey whole. Their sharp teeth are curved to hold the prey. Teeth on one side of the mouth hook into the victim. Then those on the other side are pushed forward and also hook into the victim. To swallow food, then, snakes literally pull their body over the prey. Their body stretches to take in the meal, which may be the only food eaten for weeks or months. Some snakes can even swallow a whole antelope!

Boa constrictors can swallow animals, such as this white rat, whole.

Birds: Taking to the Air

Two classes of vertebrates are made up entirely of warm-blooded animals—birds and mammals. Both have internal systems and external coverings that help them to generate and hold body heat. They also have the means to cool themselves by eliminating excess body heat. These systems are needed for warm-blooded animals to maintain a constant body temperature no matter how hot or cold the environment. Let's compare the different ways these two classes have adapted for temperature regulation and life out of the water.

Birds are vertebrates that have wings, a beak, two legs, and a body covering of feathers. Birds were the first warm-blooded vertebrates to have insulated body coverings made of feathers, adaptations for flight, hollow bones, and eggs with hard shells that allow for incubation. Most, but not all, birds fly. Birds' wings help them to fly and also insulate them. Which birds can you name that don't fly?

While mammals have hair growing from their skin, birds have feathers. Adult birds have soft, fluffy feathers next to their skin that help insulate the body. The strong, light, outer feathers, called contour feathers, are used to fly.

EVIDENCE INDICATES THAT BIRDS EVOLVED FROM SMALL MEAT-EATING DINOSAURS.

Minds On! What characteristics do birds have that enable them to fly? Picture in your mind birds that are able to fly. Think about their structure (both external and internal), their shape, their size, and other traits that make flight possible. On page 15 in your *Activity Log*, list these characteristics and compare your ideas with those of your classmates. ●

To fly and keep warm, birds must keep their feathers in good condition. To care for feathers, a bird has a special oil gland just above the base of its tail. How does the oil help? Do the activity below to find out.

See how a contour feather is adapted for flight. A shaft runs up the center of each feather. The vane on each side of the shaft is made up of many tiny branches that interlock, forming an airtight structure.

TRY THIS

Oil for Protection

What You Need
2 12-cm squares of black construction paper, petroleum jelly, dropper, water, *Activity Log* page 16

Label 1 paper square *OIL* and cover 1 of its sides with petroleum jelly. Label the other square *NO OIL*. Place 4 drops of water on each square. Describe what happens in your *Activity Log*. Now, move the drops of water around on the oily square by blowing or shaking. What happens? How do you think oil helps a bird? Explain in your *Activity Log*.

Evidence indicates that birds evolved from small meat-eating dinosaurs. Dinosaurs were a type of reptile. One such prehistoric animal, *Archaeopteryx* (är´ kē op´ tə riks), had wings and was covered with feathers. It also had bony jaws lined with teeth, claws on both feet, feathered forelimbs, and a tail.

What are the ways in which today's birds are similar to reptiles? How do they differ from reptiles? One difference is in the kind of eggs they lay. Reptile eggs have leathery shells, and bird eggs have hard shells. Another difference is in the way the eggs hatch. Cold-blooded reptiles lay their eggs and leave them. Birds, such as the great horned owl, usually lay their eggs in a nest and sit on them for days, weeks, or even months until the young have developed enough to hatch.

Fossils of **Archaeopteryx** *reveal an animal with traits of both dinosaurs and birds.*

Why do you think most birds build their nests in private, out-of-the-way places? How much do you know about owls? Let's focus on the great horned owl and its abilities.

Imagine that you are walking in a forest at night. You hear a deep, rich, almost mournful call—*hoo, hoo, hoo, hoo.* What is it? If you're quiet and observe carefully, you may catch a glimpse of the great horned owl before it flies away on powerful wings for its nightly hunt for food. This bird is an impressive example of a warm-blooded vertebrate that is perfectly adapted to life in the air. As the owl perches on a limb, its powerful four-chambered heart is beating rapidly, pumping blood to help maintain its body temperature and life processes. Well-developed lungs are bringing in oxygen. Keen eyes can easily see

Great horned owls begin to fly when they are 9 to 10 weeks old.

small and distant objects in the dark. The owl's vision is about 10 times better than a human's. It can hear much more than you can, too. Its eardrums are large, and a curve of feathers about its face scoops sound waves into its ears. Why are keen vision and hearing so vital to the life of a night hunter?

When the owl flies away, it puts into use three systems that are well adapted for this task. Its strong, light skeleton is composed of bones that are hollow and spongy inside. In fact, its skeleton weighs less than its feathers! Powerful muscles that work the wings are attached to the large breastbone. The overlapping system of smooth flight feathers helps lift and keep the bird airborne.

Mammals: The Fur Bearers

Now let's investigate mammals. You yourself are a mammal. A **mammal** is a warm-blooded vertebrate that has hair and feeds milk to its young. For what is hair used? With what other vertebrate structure that you've studied would you compare it?

Owls cannot move their eyes within the eye sockets, but instead turn their heads to watch a moving object.

Unlike other mammals, beavers continue to grow throughout most of their lives.

Platypuses are also known as duckbills and are found in Australia.

Other characteristics of mammals are mammary glands in females to feed the young, oil glands to lubricate skin and hair, and sweat glands to regulate temperature. Some mammals also have scent glands that allow them to be recognized by other similar mammals. Almost all mammals have specialized teeth and well-developed body systems.

There are three different kinds of mammals. The egg-laying mammals, monotremes, don't give birth to live young. When the young hatch, they lick milk from the skin and hair surrounding the female parent's mammary glands. Some examples of egg-laying mammals are the spiny anteater and the platypus.

A second kind of mammal is the pouched mammal, the marsupial. Pouched mammals give birth to tiny living young that crawl into the pouch of the female to develop further. Young marsupials are helpless when they're born. They have no hair and are sometimes blind. There's only one pouched mammal in North

Five-week-old opossums receive warmth and nourishment while inside the mother's pouch.

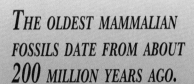

THE OLDEST MAMMALIAN FOSSILS DATE FROM ABOUT 200 MILLION YEARS AGO.

America. Can you name it? Where are most of these pouched mammals located in the world? How many different types of marsupials can you name?

The largest group of mammals is that of the placental (plə sen´ təl) mammals. With placental mammals, the young develop inside the body of the female and are well-developed when born. Which mammals can you think of that are self-sufficient when born? What are some mammals that are fairly helpless when born?

Fossil evidence indicates that mammals evolved from a group of mammal-like reptiles called therapsids (ther ap´ sid) that lived on Earth about 280 million years ago. These animals had characteristics of both reptiles and mammals. The oldest mammalian fossils date from about 200 million years ago.

Therapsids were only ten centimeters long and had skeletons similar to those of shrews. They had teeth for eating insects. It is thought that they were warm-blooded and had skin and hair.

LEARNING FROM
ANIMALS

Minds On! Make a table with three columns. In the first column, list as many vertebrates as you can think of. In the second column, classify each vertebrate as to its major group—fish, amphibian, reptile, bird, or mammal. In the third column, think of a way that each vertebrate is important. Try to think beyond human needs, and think of maintaining the balance of Earth's systems. ●

The great variety of vertebrates rounds out our picture of animal life on Earth. How do we know so much about certain animals? We're able to study these animals in their natural habitats, and so we learn more about them.

As we study animals in their natural habitats, we observe more and more of their behaviors. Some animal behaviors seem much like things humans do. For example, wolves howl in chorus and make what seems to be music. Some birds have songs that scientists have recorded and studied. Are these animal sounds really music?

Have you ever heard a wolf howling at the moon?

MELODIES IN THE AIR

For centuries, composers around the world have written music that imitates the pure, familiar notes of certain birdsongs. Composers have recorded actual birdsongs and combined them. *Symphony of the Birds* was compiled in 1955 by James Fassett, in cooperation with the Department of Ornithology (ôr nə thol′ ə jē) at Cornell University. He modified the speed of tape-recorded birdcalls to create this piece.

When birds sing, are they making music or are they simply trilling sounds that were programmed into them at birth? Studies have shown that only male birds sing, and they do so to send a warning to other males or an invitation to a mate. They learn to perfect their songs as youngsters, in much the same way human babies learn to talk. Baby birds sing a simple song, and deaf birds never learn to sing their species' song correctly.

What's more, studies have shown that birds can learn to distinguish between different styles of human music. They have a tremendous ability to recognize the sounds of their own songs. This ability also seems to extend to the sounds of our music.

What we may consider to be music or singing is the way some animals communicate. Conduct research to find out what scientists have discovered about the communication methods of animals. Write a short essay about how a certain animal communicates, and present it to the class.

SUM IT UP

Vertebrates, being complex animals, have retained some characteristics of earlier, simpler organisms including invertebrates. In your observations of the five classes of vertebrates, you were able to recognize these traits and determine how complex vertebrates might have evolved from simple organisms. The more we know about how animals are similar and how they differ, the more we can learn about ourselves and all life on Earth.

CRITICAL THINKING

1. What might be one advantage of being cold blooded?

2. How does the swim bladder of a bony fish save energy for the fish?

3. In what ways are the eggs of birds more protected than the eggs of cold-blooded vertebrates?

4. Placental mammals are more numerous than marsupials or monotremes. Suggest a reason why.

5. Suggest a reason why mammals have a smaller number of young than other animals.

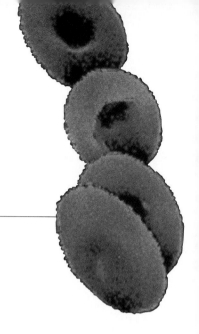

How Do Body Systems

Maintain Life?

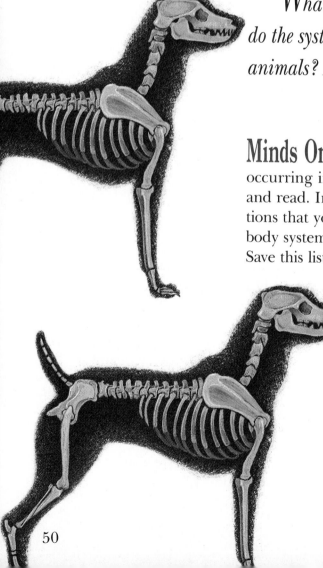

What makes an animal's body function? How do the systems in your body differ from those of other animals? How are they similar?

Minds On! What enables animals to perform so many functions? Think about what's occurring inside your body now as you sit at your desk and read. In your *Activity Log* on page 17, list functions that your body is carrying out and the name of the body system you think is responsible for each function. Save this list. Discuss your ideas with the class. ●

Animals have similar organ systems. In this lesson, you'll look at respiratory, digestive, circulatory, excretory, skeletal, muscular, and nervous systems. The differences in these systems are in part a result of adaptations to different environments. However, as you study, you'll see that the systems of different animals are

Did you ever think of the skeleton as a body system?

50

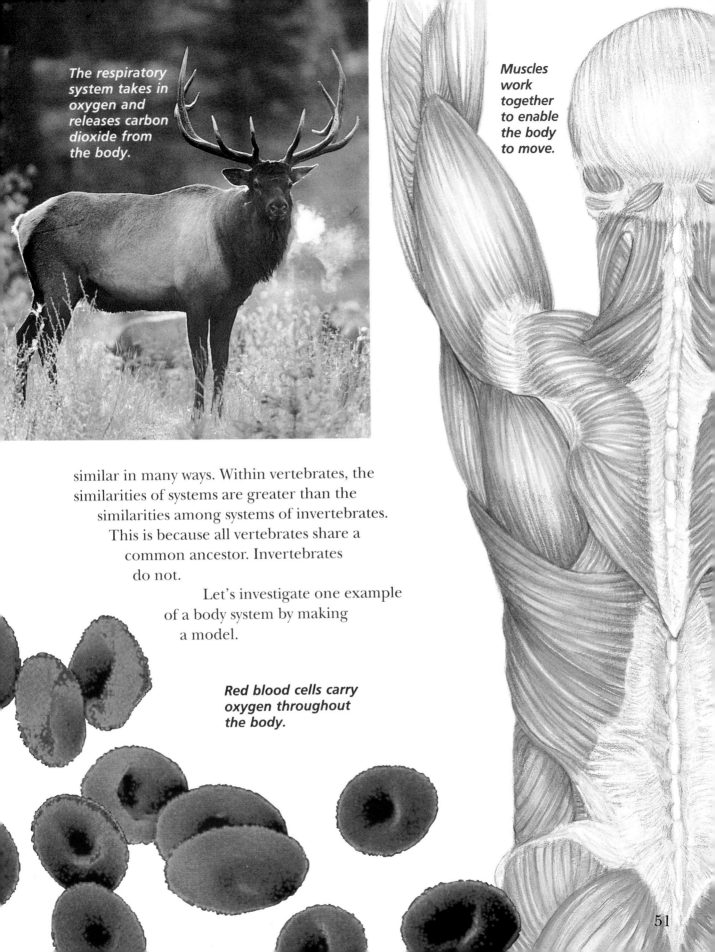

The respiratory system takes in oxygen and releases carbon dioxide from the body.

Muscles work together to enable the body to move.

similar in many ways. Within vertebrates, the similarities of systems are greater than the similarities among systems of invertebrates. This is because all vertebrates share a common ancestor. Invertebrates do not.

Let's investigate one example of a body system by making a model.

Red blood cells carry oxygen throughout the body.

51

Activity!

How Does Breathing Occur in Mammals?

One body function you may have just noted in your **Activity Log** is breathing through the use of your respiratory system. What is breathing? If you've ever tried to hold your breath, you know that you can't do it for very long. In this activity, you'll build a model of the respiratory system of a mammal. You can discover how the different parts of the respiratory system work together.

What You Need

2-L plastic drink bottle with
    bottom cut out
Y-shaped plastic tube
rubber stopper
2 small balloons
plastic wrap
3 small rubber bands
tape
goggles
Activity Log pages 18 - 19

What To Do

1 Use the rubber bands to fasten the balloons to the Y-shaped tube. *Safety Tip*: Be sure to wear goggles when working with rubber bands.

2 Insert the top of the Y-shaped plastic tube into the hole in the wide end of the rubber stopper.

Safety!

See the *Safety Tip* in Step 1.

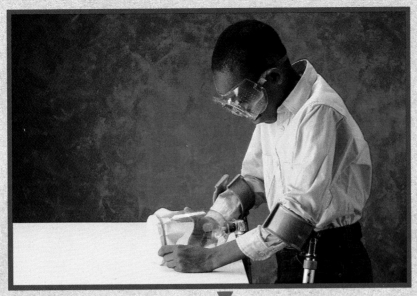

3 Reach inside the bottle and insert the rubber stopper into the bottle's neck.

4 Stretch plastic wrap over the bottom of the bottle. Use a rubber band to fasten it tightly. Put a small loop of tape on the plastic wrap's outside surface.

5 Push up gently on the plastic wrap. Observe the balloons inside the bottle.

6 Pull down gently on the plastic wrap. Observe the balloons inside the bottle.

What Happened?

1. What happened to the balloons inside the bottle when you pushed up on the stretched plastic wrap?
2. What happened to the balloons inside the bottle when you pulled down on the stretched plastic wrap?

What Now?

1. What would happen if the stretched plastic wrap weren't flexible and couldn't be pulled up or pushed down?
2. Explain how pushing or pulling on the stretched plastic wrap affects the balloons inside the bottle.
3. How does this experiment model the process of breathing in mammals?

EXPLORE

How BODY SYSTEMS Function

Y ou've just seen how breathing works within your body. This is one function of one of the several systems in your body. What exactly is a body system? Body systems are groups of organs that perform a specific function and contribute to sustaining the life of an organism. You'll examine each body system in detail and discover how each contributes to keeping animals alive.

THE RESPIRATORY SYSTEM

A **respiratory system** is a system of organs that help an animal exchange oxygen and carbon dioxide with the environment. Mammals, birds, reptiles, and amphibians all have lungs. Most fish have gills. For an example of how gills function, examine the diagrams on this page.

How does the air enter and leave the lungs of mammals? A sheet-like muscle at the bottom of the chest, called the **diaphragm** (dī′ə fram′), separates the chest cavity from the

As water flows over the gills, the dissolved oxygen diffuses into the animal's blood. At the same time, carbon dioxide from the fish's blood diffuses into the water passing over the gills. Gills are located behind the head in slit-like openings.

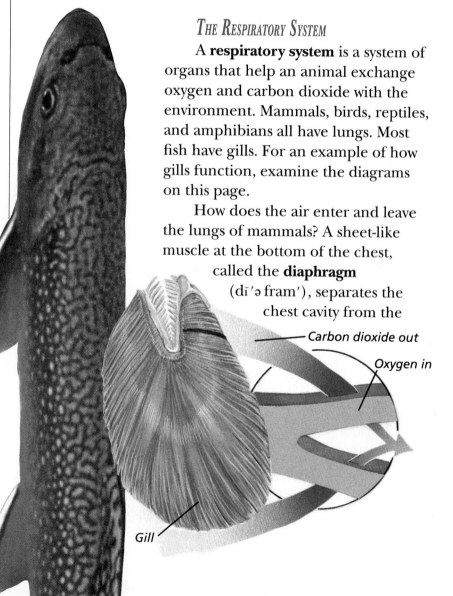

Carbon dioxide out

Oxygen in

Gill

abdominal cavity. In the Explore Activity, you determined that the diaphragm allowed air to enter the lungs when it contracted. When it relaxed, air rushed out.

Once air enters the lungs, what happens? The lungs of mammals have millions of microscopic sacs that allow large amounts of oxygen and carbon dioxide to move in and out of the blood. Once in the blood supply, oxygen is used by the cells. Oxygen combines with sugars obtained from food that has been digested. Energy is then released for the body to use to move and grow, and carbon dioxide waste products and water are formed.

Lungs

Air sacs

In order for a mammal's respiratory system to function properly, there must be a large, moist surface area over which air passes, such as the air sacs. Let's compare some respiratory systems of different animals.

Some invertebrates such as sponges, cnidarians, and flatworms have no specific organs or tissues for breathing. Their body cells didn't evolve structures to breathe. Their body cells are in constant contact with water from which they receive oxygen. Wastes such as carbon dioxide are released directly into the water.

Minds On! Other animals, such as earthworms, exchange oxygen and carbon dioxide through their skin. Insects, however, have tiny openings in their exoskeletons through which gases are exchanged. Think about the environment of earthworms and insects. How is the respiratory system of each animal adapted for its environment? Write your ideas in your *Activity Log* on page 20. ●

Air sacs connected to a bird's lungs serve to cool the bird's blood and aid in oxygen exchange.

Gizzard

Crop

Intestine

Digestive system of a rooster

THE DIGESTIVE SYSTEM

The **digestive system** is a group of organs that breaks down the food animals take in and changes it to a form cells can use. To see how a bird's digestive system works, do the Try This Activity.

After swallowing hard materials like seeds, birds must digest them. Because these materials are difficult to digest unless ground up, birds have organs called gizzards. The grit the bird has swallowed is stored in the gizzard, where it acts to grind the seeds much as you ground seeds with gravel in the Try This Activity.

Some animals, such as tapeworms, have no digestive system. They live inside the digestive systems of other animals, and food that is already broken down passes into their bodies by diffusion.

The digestive systems of animals have either one or two openings for food to enter and waste to leave. Sponges, cnidarians, and flatworms have only one opening by which food enters and wastes leave the body. Animals like earthworms, fish, and humans have two openings—one for food to enter the body called the mouth, and another for wastes to leave the body called the anus.

TRY THIS

Activity!

How Do Birds Get Nutrients From Food?

In what ways do birds digest their food? With this model, you can find out.

What You Need

self-sealing plastic bag, 1/2 c of cracked corn or sunflower seeds, 1/2 c of gravel, *Activity Log* page 21

Place some cracked corn or sunflower seeds into the bag. Seal the bag so that no seeds can drop out. Use your hands to grind the seeds within the bag. What happens to the seeds?

Many birds swallow grit, small pebbles, egg-shells, and other hard materials. Open the bag and place some gravel inside. Again, grind the contents of the bag with your hands. Did you note any differences in the seeds after you added the gravel? What function do you think hard materials such as gravel perform in the digestive system of a bird? Write your observations and ideas in your *Activity Log.*

Minds On! How many parts of your digestive system can you name? On page 22 in your *Activity Log,* record as many parts of your digestive system as you can. Compare your list with your classmates. ●

Do all animals have the body parts you listed in the above Minds On? Earthworms differ from humans because they have a crop and a gizzard. Food is taken in through the mouth and stored in the crop. Food is then moved to the gizzard where it's ground up.

Other animals, such as fish, are similar to humans because they have teeth and tongues. Fish are different from humans, though, because they can't move their tongues. Humans and fish both have stomachs where food is digested.

Digested food then moves to the intestines in earthworms, fish, and humans. This food has been changed physically and chemically so that its nutrients can be absorbed. Not all of the digested food is usable, though. The unusable remains are excreted, or forced out of the body.

Different kinds of animals eat different foods. Their bodies have organs that have evolved to adapt to their environment and food supply. Animals that eat only plants are called **herbivores.** Herbivores, such as cows and koalas, eat large amounts of plants. They have teeth like the ones shown and must eat often. Plants don't contain high levels of protein, so large amounts must be eaten for herbivores to remain healthy.

Animals that eat only meat are called **carnivores.** What carnivores can you name? What differences do you see between the teeth of herbivores and those of carnivores?

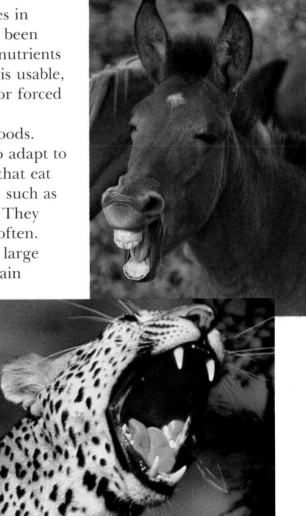

The teeth of herbivores like this mule often grow throughout the lifetime of the animal.

The teeth of carnivores are usually sharp and long to cut and tear the meat they consume.

Some animals eat both plants and meat and are called **omnivores.** Raccoons, opossums, and many humans are some examples of omnivores.

THE CIRCULATORY SYSTEM

Each cell in the body needs nutrients to survive. How do the cells receive the nutrients once they are absorbed during digestion? The blood delivers materials such as oxygen, water, and nutrients to the cells and removes wastes like carbon dioxide from the cells. The blood, blood vessels, and heart make up the **circulatory system**. This system transports materials to and from the cells of the body.

Not all animals have circulatory systems. Animals like sponges, cnidarians, and flatworms use the water in which they live to transport materials to their cells.

If a circulatory system functions so that the blood is not enclosed in vessels, it's called an open circulatory system.

The water in which these corals and sponges live works like blood in other animals.

CAPILLARIES ARE ONLY ABOUT AS WIDE AS ONE CELL.

Some animals that have open circulatory systems are grasshoppers, crayfish, and spiders. Vertebrates have closed circulatory systems in which the blood moves inside blood vessels.

In vertebrates, the heart pumps blood through blood vessels that reach to all parts of the body. Hearts of vertebrates are divided into chambers. The hearts of fish have two chambers, the hearts of most reptiles and amphibians have three chambers, and those of birds and mammals have four chambers.

There are three types of blood vessels in vertebrates. Arteries contain blood that flows away from the heart to the cells. Veins contain blood that flows back to the heart from the cells. Capillaries are the connections between arteries and veins.

Capillaries are only about as wide as one cell. Why must they be so small? In these narrow passageways, the blood makes contact with the cells of the body. It is here that digested food passes through the capillary wall and into the cell.

Blood not only carries oxygen and food to cells, but it takes carbon dioxide waste away. This gas is taken to the lungs and removed from the body during exhaling.

Grasshoppers have open circulatory systems. Blood moves through a grasshopper's body without traveling inside vessels. The heart pumps blood into a single large vessel. This vessel then empties blood into the body spaces.

Fish Heart

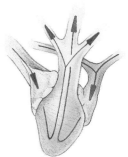

Amphibian Heart

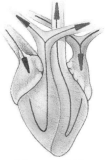

Mammal Heart

Vertebrate hearts are divided into chambers. Amphibians have 3 chambers, fish have 2 chambers, and mammals have 4 chambers.

THE EXCRETORY SYSTEM

Other wastes are removed from the body by the **excretory system.** Organs such as the skin and kidneys help the body get rid of liquid waste materials. Chemical wastes are carried by the blood and must be removed from the body to keep the cells and tissues healthy. To find out how this process works, do the Try This Activity below.

Many animals don't have kidneys at all. In sponges and cnidarians, wastes pass out of the cells directly into the water in which they live. Earthworms excrete wastes through a pair of tubes in each body segment that connect the inside of the animal to the outside.

When body proteins are broken down, a poisonous waste called urea (yù rē´ ə) forms. The blood picks up urea from cells and carries it to the kidneys for removal from the body.

Artery

Vein

Urea is removed from body

Kidney

TRY THIS

Activity!

How Do Kidneys Work?

You can investigate how the kidneys help to clean the blood of wastes.

What You Need

2 plastic containers, soil, plant material, water, filter paper, *Activity Log* page 23

Mix soil and plant material with water in 1 container. Pour some of the water from one container into a second container lined with the filter paper. Examine the water after it has run through the filter paper. How is the water different from the original mixture? Examine the filter paper. What does a filter do? How does this compare to the function of kidneys? What might happen if the kidneys stopped functioning? Write your observations and answers in your *Activity Log.*

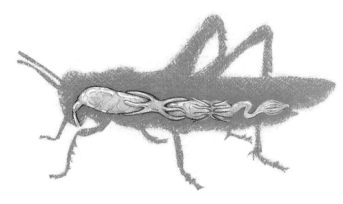

The excretory system of insects consists of tubular structures that absorb body wastes and change them into crystals. These crystals pass into the intestine and out of the body as wastes. In vertebrates, kidneys filter wastes from the blood and help regulate the salts and water in the blood.

THE SKELETAL SYSTEM

In Lesson 2, you discovered another important system. The **skeletal system** provides a framework that supports and protects the body and helps the animal to move. What animals can you think of that have soft bodies with little support? Other animals have shells secreted by the animals themselves. Mollusks have shells to protect them from their environment. Other animals such as grasshoppers, frogs, and humans have skeletons.

Minds On! Some animals have skeletal systems on the outsides of their bodies. What is this type of skeletal system called? What are some advantages and disadvantages for animals with this type of skeletal system? Write your answers in your *Activity Log* on page 24. ●

Beetles are the most common type of insect. There are approximately 300,000 different kinds.

In mammals, fins and flippers are supported by bones and cartilage.

Most vertebrates have endoskeletons made of bone. **Bone** is a hard substance that contains the minerals calcium and phosphorous. Skeletal tissue is made up not only of bone but also of cartilage. Cartilage is a tough, elastic material that is more flexible than bone. It's found in the ears, nose, and at the ends of bones where it acts as a cushion.

Vertebrates have different shapes, but most vertebrate skeletons are very similar. Each skeleton has a backbone with two pairs of limbs (arms, legs, wings, fins, flippers). These limbs are quite similar in structure and indicate derivation from a common ancestor. They also serve a similar function. Investigate bones in the Try This Activity.

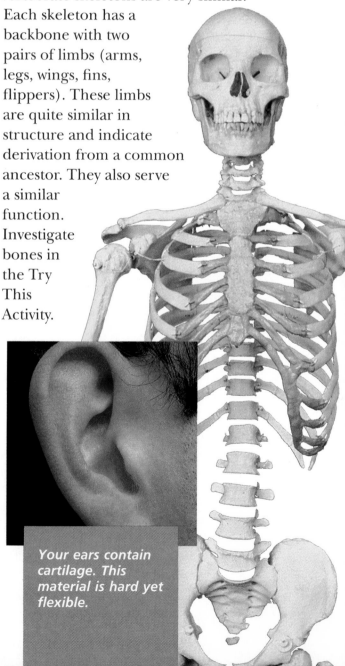

Your ears contain cartilage. This material is hard yet flexible.

TRY THIS

Activity!

Two Bones To Pick

You can investigate how bird bones are similar to and different from mammal bones.

What You Need

2 bird bones (1 whole, one cut across)
2 mammal bones (1 whole, one cut across)
hand lens
Activity Log page 25

Compare bones from a bird and a mammal. Use a hand lens to study the inside of each bone. Sketch and label a cross section of the bird bone and the mammal bone in your *Activity Log.* Note any similarities in structure.

Hold both of the whole bones. Compare their shape, weight, and texture. Which has more mass? Which seems stronger? Sketch and label each whole bone in your *Activity Log* and list the similarities and differences between them. Think about the way each of these bones is used in the living animal. How do the uses of each bone explain the similarities and differences? *Safety Tip:* When you're finished working with the bones, wash your hands with soap and water.

MUSCULAR AND NERVOUS SYSTEM

The skeletal system has another important function—it provides areas of attachment for the muscles. Vertebrates use muscles to move their bodies. The **muscular system** contains muscle tissues that act together to allow for movement of the animal. How does this work?

Bones move by the action of paired muscles that react to nerve impulses by shortening or lengthening. Each time one muscle shortens, the other muscle in the pair lengthens.

An animal requires the ability to react to its environment in order to survive. A nervous system helps an animal detect and respond to changes in the environment. A **nervous system** is made up of nerve cells, sense organs and, usually, a brain.

The simplest animals that have nervous systems are the cnidarians. Their nervous system consists basically of a nerve net extending throughout the body. Flatworms have a more complicated nervous system with eyespots, ganglia (gang′ glē ə), and nerve cords. Because of their more advanced structures, flatworms have a greater variety of responses to their environments than cnidarians.

The nervous system of insects has evolved into a more complex structure with antennae, a brain, a nerve cord, and nerves throughout the body. Insects like grasshoppers can respond even more quickly to their environments than flatworms.

All vertebrates have a central nervous system, which includes a brain and spinal cord. These organs control all body activities. Imagine a plate of food in front of you. Your central nervous system supplies the sensations of smell and taste, the use of your hands to pick up a fork, the working of your stomach, even the conscious

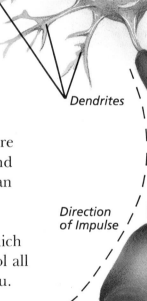

Nerve cells have fibers that reach out to all areas of the body to control and regulate body processes. One kind of branching fiber, the **dendrite**, receives nerve impulses. Another kind, the **axon**, sends nerve impulses on to the next nerve cell. Bundles of dendrites and axons make up the actual nerves.

Nucleus

Dendrites

Direction of Impulse

Axon

thought that you're full. The vast network of nerves controlled by the central nervous system reaches every part of the body to control senses, organ functions, movements and thoughts.

How does one nerve cell relay the message to the next across this gap? Scientists have learned that the nerves transmit their messages electrically by using sensitive electrical equipment that measures tiny electrical currents. A message, called an impulse, is sent along the nerve cell in the form of electrical energy. The impulse passes quickly along the fibers—although its rate of speed is only about 100 meters (300 feet) per second compared to a regular electrical current's speed of 300,000 kilometers (186,000 miles) per second. However, the impulse meets a gap when it reaches the branched endings of the cell. How does the message cross this space? An electrical impulse releases a chemical at the synapse. This chemical crosses the gap and stimulates the next cell, which carries the message electrically. The chemical transmitter could be thought of as a boat carrying a message from a runner on one side of a river to a runner waiting on the other side.

DOES THIS MEAN THAT YOUR BODY IS LIKE AN ELECTRIC APPLIANCE?

Does this mean that your body is like an electric appliance? Not exactly. Both nerve impulses and electric currents travel in one direction and produce a measurable voltage. However, electrical currents pass along a wire and are generated by an outside force such as a battery or generator. Nerve impulses travel along a living fiber and undergo an electrochemical change in order to keep the message moving.

Structures on each end of nerve fibers are the means by which the nerve cells make connections with each other. Since nerve cells must send messages to regulate body processes, this point of contact is very important. A tiny space called a **synapse** *separates one nerve cell from the next.*

Nerve cell

Synapse

Nerve cell

Sense organs are also a part of many animals' nervous systems. Insects have eyes and antennae that detect odors and sense their environments. Insects also have an organ that they use to hear sounds. What senses do you have that allow you to respond to your environment?

Directed by the nervous system, all these body systems work together to meet an organism's needs for life.

- They provide food and oxygen for cell activities.
- They eliminate wastes after digestion.
- They support and protect the organs and allow the animal to move.
- They allow the organism to sense and respond to its world.
- They work together to maintain life.

Certain chemical substances can cause your body systems to function differently. Some of the effects are helpful, but other effects can be harmful.

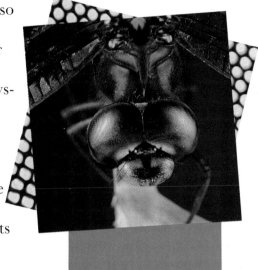

The eyes of most insects contain several lenses to sense their environment.

Health  Link

SYSTEMS, CELLS, AND DRUGS

How can drugs affect your body systems? A drug is a chemical or mixture of chemicals that affects the body when it is introduced. Drugs work by interacting with body systems and changing their functioning or by attacking invading microorganisms.

The effects of a given drug may be helpful or harmful to the body. Many drugs stimulate certain types of cells and increase cell activity. Others depress certain types of cells and lower cell activity. Alcohol and tranquilizers are two drugs that slow the transmission of messages along the nerve cells. How can this action have harmful effects?

Aspirin is a drug that acts on body systems in several ways. It blocks the production of pain impulses in the nervous system by stopping the formation of substances that make pain receptors sensitive. Aspirin also lowers body temperature when illness causes a fever. More blood then flows to the skin where it can release more heat. Aspirin is the most frequently used drug in the United States today. When do you think it could be harmful?

Minds On! Return to your list in your *Activity Log* on page 17 — your list of the functions that occur inside your body as you sit at your desk. How accurate were you when you listed the body systems responsible for these actions? Now that you've studied the systems of your body, try again to list what is going on inside your body and name the system responsible for these functions in your *Activity Log* on page 17. ●

Body systems work together to maintain life. All animals have systems to carry out life functions. They must also reproduce themselves, grow, and develop. You'll discover how animals do this in the following lessons.

TECHNOLOGY AND
BODY SYSTEMS

Every living organism must carry on certain processes that maintain and control life. All animals are composed of cells, and their systems are created from many types of specialized cells. What very specialized cells help some animals sense their environments? Read the Literature Link on the next page to find out how animals sense their environment through vision.

Specialized cells make up these magnified hair follicles.

DEAR BRONX ZOO

"**How do mammals find and catch their food?**

. . .Most carnivores have keen eyesight and a good sense of smell to help them track down prey. They have what's called **binocular vision**, which means their two eyes face forward. Binocular vision helps an animal hunt by giving it a sense of **depth perception**, enabling it to see three dimensionally and allowing it to focus both eyes on the same object. Herbivorous animals, on the other hand, generally have **monocular vision**: Their eyes are set far apart on either side of their head so that they can see in two directions at once."

In *Dear Bronx Zoo,* authors Joyce Alterman and Sue Goldberg answer thousands of questions about animals. What would you like to know about animals? Come up with three questions. Exchange questions with a classmate and do research to answer them. If libraries don't have what you need, perhaps the staff of a nearby zoo can offer research help.

Most animals do have many specialized cells and tissues that act together in organs and organ systems to help the animal sense and control its environment. Each system is specialized to carry out its tasks, but all systems are dependent upon one another. At times, though, some systems fail to work properly. What happens then? Turn the page to find out.

Owls eyes are especially adapted for hunting prey.

NEW PARTS FOR OLD—NEW LIFE, NEW QUESTIONS

The failure of a major organ such as a kidney, liver, or heart often means death for the organism. For centuries, doctors relied on repair or removal of damaged organs to save lives. If these procedures didn't work, the patient died.

Thanks to medical technology of the past three decades, now there is another alternative. Many diseased organs or tissues can be replaced with healthy ones. The process of organ transplantation means that thousands of people each year are given a second chance for life.

Transplant surgery can restore the function of a body system to almost normal again. What exactly is transplant surgery, and how is it done?

Usually, an organ is removed from one person and put into another. Sometimes, the surgeon may simply take tissue from within an individual's body and reposition it. The organs most frequently transplanted are the cornea (the clear front portion of the eye), heart, liver, kidney, pancreas, skin, and bones or portions of bones.

Some organs, such as kidneys, are paired, and the body can function with just one. This gives individuals the ability to donate one of their paired organs. Other organs can be removed from donors who have died. Often organ donors are family members because their tissues are more likely to be compatible, or similar, to that of the receiver. Compatibility is important, since each body has an immune system that attacks foreign objects such as bacteria and viruses. When an organ is transplanted into a body, the new body's immune system may think that the transplanted organ is "foreign" and try to eliminate it. Rejection of new organs is a major problem in organ transplants.

The donation of a life-giving organ is an important gift. Now that technology provides the ability to transplant organs, some organ-transplant procedures have become common. But there are always more people who need transplant operations than there are organ donors. The organs to be transplanted must be healthy.

The ability to transplant organs has led to some difficult questions for medical science today. For example, who receives a donated organ? For one heart, there may be several patients waiting. What are the criteria for deciding, and who decides?

Minds On! Think about how technology has changed the way our body systems function. Brainstorm in a small groups to determine ways advanced technology has enabled people to live longer. Think about each body system, and record ways each system has been helped through advances in medicine in your *Activity Log* on page 26. Discuss these ideas with your class. Do you see any disadvantages to using these techniques to help people? If so, what are the problems? What do you think can be done to minimize them? ●

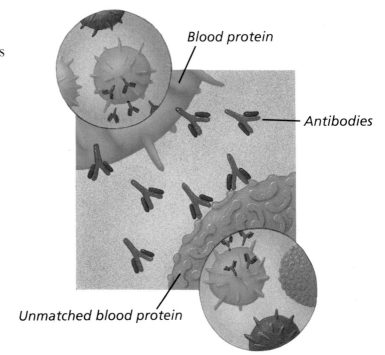

Blood protein

Antibodies

Unmatched blood protein

One way to prevent rejection of donated organs is to match the proteins of the donor and the receiver as closely as possible. The closest match would be between identical twins, who have exactly the same proteins in their blood. If the blood proteins are identical, the transplanted organ can be accepted by the body. If the blood proteins differ between the donor and recipient, then antigens form in the recipient's blood and the new organ is rejected.

Sum It Up

You've discovered how body systems work in various kinds of animals. Even though each system has separate and distinct functions, all systems depend upon each other to sustain life. When you explored the respiratory system of mammals, you were able to discover how the organs work together to enable the system to function. You determined that each system is related to the others, and that all act together to support the organism. When all systems of an organism are functioning well, then the animal is in good health.

Critical Thinking

1. What advantage does a closed circulatory system have over an open circulatory system?

2. Many water animals have no skeletons. How are their bodies supported?

3. Why is it important for vertebrates to have foods that contain calcium?

4. Suggest a reason why the combined surface area of the tiny sacs in the lungs needs to be very large.

5. Why is it important that the kidneys function properly?

REPRODUCTION, GROWTH, AND

DEVELOPMENT

Animals have existed on Earth for millions of years. What has allowed animals to survive for such a long period of time through all of Earth's changes?

How do animals continue to exist on Earth? Consider the images on these two pages. A mass of jelly-like frog eggs shimmers in the shallow water at the edge of a pond. Newly-hatched baby turtles return to the sea. A human baby wakes up hungry and cries. What do these pictures have in common? They're all evidence of an ongoing life cycle in which all animals participate.

Not only must animals maintain life processes to survive, they must also reproduce if their species are to survive. Each generation must produce offspring. The young must grow and develop into adults who can in turn produce more offspring.

Minds On! The study of the life cycle in various animals is also a study in a wide range of forms and adaptations that all animals have evolved. While you examine the photographs, think about the following questions.

● Do all animals have two parents?
● Do all animals care for their young?
● How many kinds of animals hatch from eggs?
● Do the offspring of all animals look like the parents?

Write your answers to these questions in your *Activity Log* on page 27 and then discuss your answers with the class. To examine one animal's life cycle, turn the page. ●

Activity!

How Do Butterflies Form?

What is a chrysalis (kris'ə lis)? How did it form? What is happening to the larva within that protective covering? A kitten or puppy grows by increasing in size until it becomes an adult. Its body form doesn't alter significantly. However, some animals grow and develop in stages, and each stage looks very different from the one before it.

In this activity, you'll observe and record the changes in a butterfly chrysalis to discover how one type of animal grows and develops.

What You Need

Painted Lady butterfly larvae
container to hold larvae
hand lens
Activity Log **pages 28-29**

What To Do

1 *Safety Tip:* Be very careful with the larvae. Do not allow them to be crushed or damaged. Place the larvae into the container. Observe the larvae. Record their color and texture in the ***Activity Log***.
Safety Tip: Be sure to wash your hands with soap and water after handling the larvae.

Safety!

See the *Safety Tips* in step 1.

2 Observe the larvae daily as they change form. What changes do you see? Be sure to record your information in your *Activity Log.* Predict what changes are occurring within the chrysalis. Draw what you think the larvae look like within each chrysalis. What's going to happen to the larvae?

3 When the butterfly emerges, draw the butterfly in your *Activity Log.* Note the color and any patterns on the wings.

What Happened?

1. How long did the butterfly take to form? Describe what occurred within the chrysalis. Compare your notes with the notes of your classmates.
2. What stages did the butterfly go through in its development?

What Now?

1. What other animals can you think of that go through similar changes while they are growing or developing?
2. How does this method of growth differ from the way humans grow and develop?
3. During which time of the year do you think the different stages of the butterfly normally occur? Why?

METHODS OF REPRODUCTION

As you discovered in the Explore Activity, a butterfly doesn't have the same form throughout its life. It's one of a number of animals that go through several stages, each with a different form, as it grows and develops to adulthood. From the outside, it doesn't appear that much is happening to the animal. However, inside the chrysalis, incredible changes are occurring in the larva. Once the changes in the larva have occurred, an adult butterfly emerges from the chrysalis looking very different.

The process by which some animals go through very different stages of growth and development to reach adulthood is known as **metamorphosis** (met´ ə môr´ f ə sis). What other animals do you know of that go through metamorphosis? We'll look more closely at this process later in this lesson.

All animals begin with the process of reproduction. Where did the butterfly begin? Where did you begin?

Just as life processes go on within each cell, the process of reproduction begins with the cell. All cells are reproduced from other cells. Cells can split and form two new cells.

When a cell divides by mitosis, genetic material is duplicated. The material lines up in organized strands. Each strand carries a certain genetic code that holds the traits and structures that the cell should have. Each new cell receives one complete copy of the genetic code.

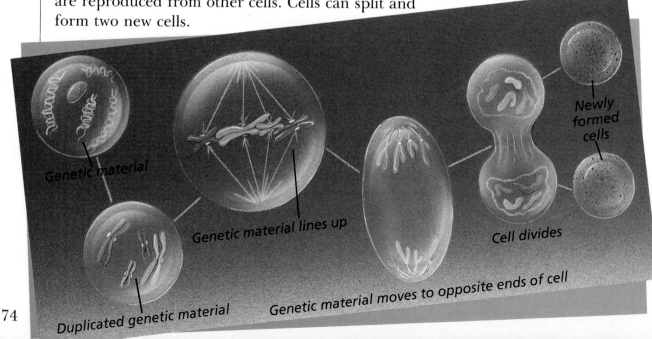

Genetic material

Duplicated genetic material

Genetic material lines up

Genetic material moves to opposite ends of cell

Cell divides

Newly formed cells

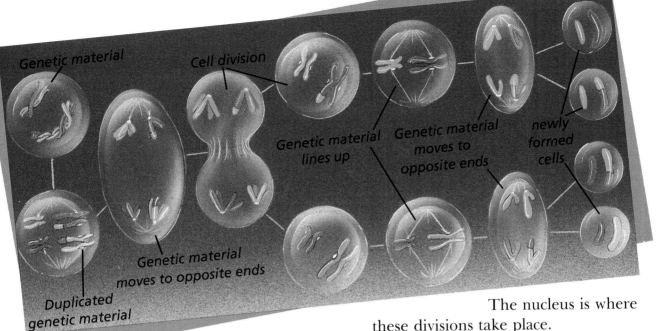

Genetic material

Cell division

Genetic material lines up

Genetic material moves to opposite ends

newly formed cells

Genetic material moves to opposite ends

Duplicated genetic material

Meiosis allows for the formation of egg and sperm cells. The genetic material lines up into strands. During division, one-half of each pair of strands goes to the newly formed egg or sperm cells. This allows for a new organism to have traits from both parents when egg and sperm cells combine.

The nucleus is where these divisions take place.

Body cells reproduce by a method called mitosis. **Mitosis** (mī tō´ sis) involves the duplication and the equal division of the cell's genetic material into two new, identical cells. Mitosis is the process by which an organism produces more cells in order to grow.

Minds On! Most cells reproduce themselves by mitosis. This allows the organism to grow and to replace cells that have died. As each cell divides to form two, the process makes certain both have a copy of the genetic code. Why do you think this is necessary? Write your explanation in your *Activity Log* on page 30. What would happen if the cell divided the codes at random so that each new cell had only a part of the code? ●

Most animals have a second process of cell division in which sex cells reproduce themselves. **Meiosis** (mī ō´ sis) occurs when certain cells in reproductive tissue divide so that each contains exactly half of the hereditary material. The strands that carry genetic codes occur in pairs. During meiosis, one half of each pair goes to the new cell. The resulting new cells each have one strand from the original pair.

In male animals, meiosis results in the formation of sperm cells. This takes place in organs called the testes (tes´ tēz). In females, meiosis results in the formation of egg cells in the ovaries (ō´ və rēz). Each of these types of cells contains half of the normal number of strands with genetic material. Why half?

REPRODUCTION

All animals reproduce in one of two ways. Some animals can produce new offspring by asexual reproduction. **Asexual reproduction** is the production of offspring from one parent cell. There are no egg and sperm cells involved. The cell from the parent undergoes mitosis and forms an offspring identical to the parent. Some animals that reproduce by asexual reproduction are sponges, cnidarians, and flatworms.

Some simple animals, like the hydra shown on this page, may reproduce asexually by budding. The new animal begins as a knob on the side of the parent. The knob grows and forms tentacles. Then it breaks off, having become an identical but smaller version of the adult.

Some animals can reproduce by regeneration. **Regeneration** occurs when an organism separates into two or more parts and each part forms a new organism. If you've visited the seashore, you may have seen a starfish like the one shown here with one small ray and four large ones. It was probably regenerating a lost ray.

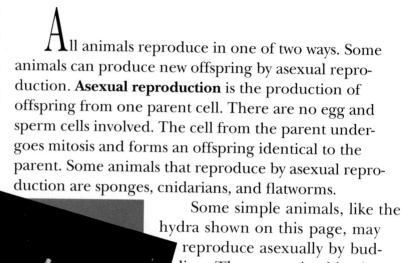

A hydra forms a bud that is identical to the parent.

A single ray of a starfish may break off. This ray will develop into another starfish while the original starfish will regenerate a new ray.

Puppies can have the coloring of one parent or both parents.

Elsewhere, the severed ray was becoming a complete starfish as well through the process of regeneration.

Other animals reproduce sexually. **Sexual reproduction** occurs when offspring are produced from the combination of the material found in the nucleus of two different parents. **Fertilization** occurs when a sperm cell combines with an egg cell. This new cell, called a **zygote** (zī´ gōt), is the basis for a new organism. The zygote grows and divides rapidly by mitosis. Because it has genetic material from each parent, the zygote combines the genetic codes in new ways. You've probably noticed that the young of an animal have the same general structure and traits as the parents, but they're never exactly the same, like the dogs pictured here.

In effect, meiosis is like shuffling a deck of genetic material. When the "cards" or genetic codes are "dealt" to the offspring, different combinations occur. Each set of codes contains information for specific traits. The codes that direct the formation of the eyes in animals may direct that the eyes be colored blue, green, black, brown, or hazel.

Sexual reproduction means not only that a species continues, but that the species has offspring that have different traits from the parents. How do you think this variety helps a species to adapt and survive in a changing world?

Blue eyes

Hazel eyes

Brown eyes

The zygote divides, forming two cells. These two cells form four, the four divide into eight, and so on. Soon a hollow ball of cells develops. This process is called cleavage (klē′ vij). It's in this process that an embryo (em′ brē ō′), an early stage of the developing organism, forms.

For a species to survive, the zygote must be able to grow and develop safely. Some animals have adapted to their specific environments so that fertilization occurs outside the female. For example, a female fish releases her eggs into the water. In many instances, the male swims behind and just above her, releasing sperm at the same time. In other animals fertilization occurs internally, inside the female's body. This requires the male and female to mate, or join, in order to unite the sperm and egg cells. For example, the male grasshopper deposits sperm within the female's body, where her eggs are fertilized.

Minds On! For fertilization to occur, the sperm need a liquid in which to swim to the eggs. Why do you think animals that live in water use external fertilization? What disadvantage do you see for water dwellers that fertilize eggs externally? How could they overcome that disadvantage? Which means of reproducing do you think evolved first, sexual or asexual? Write your ideas in your *Activity Log* on page 31.●

All organisms begin from a single cell. How does one cell of the embryo begin producing liver cells, while another produces nerve cells and another bone cells? Scientists are still studying this question.

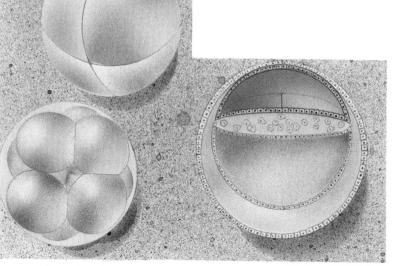

These cells continue dividing until a three-layered ball has formed. At this point, tissues can develop. The simplest animals have only two layers of cells. An embryo has three layers, so it's already more complex than the simplest animals. Each of the embryo's tissue layers will develop different structures as the organism continues to develop.

Now that the developing embryo has three different layers of tissue, what happens? Specific organs and systems begin to form. For example, on the top of the embryo, certain cells divide quickly and form a flat plate. With more divisions, the sides of the plate form two folds that extend the length of the embryo. Gradually, the folds fuse together and form a tube. This structure develops into the brain and spinal cord.

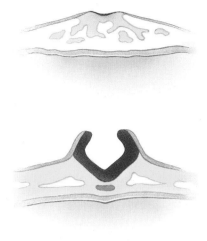

Somehow the genetic code carries information that directs the production of certain enzymes. The enzymes cause chemical reactions that in turn stimulate the cells to become each particular type. Some genes may have *on* and *off* switches that operate during development. Genes that will cause production of the right enzyme are switched on at just the right stage of development.

Minds On! At certain stages of their development, the embryos of a number of different vertebrates look almost the same. Study the embryos of these different animals. Evidence shows that animals with similar embryos and similar stages of development shared a common ancestor at some point in their evolution. The more the embryos resemble each other, the closer the relationship is thought to be. Why might this similarity in development show a common ancestor? What do these photographs suggest about the evolution of vertebrates? Write your ideas in your *Activity Log* on page 32. ●

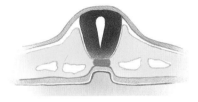

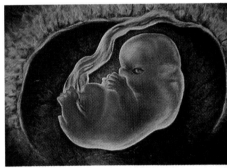

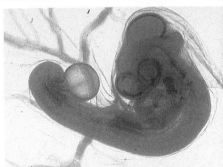

The embryos of certain animals resemble each other at certain stages of development.

For animals that develop inside the mother, the period of growth before birth is called **gestation** (jes tā´ shən), or pregnancy. While the tissues, organs, and body systems are developing, the embryo must receive food and oxygen and excrete wastes. It must also be protected. What keeps the embryo alive? Examine the diagrams on the following page to discover how some mammals keep their embryos alive.

The diagram you just studied helps to explain how mammals provide for their developing, unborn young. What about animals that don't develop inside the mother, but instead develop inside an egg? In addition to a sac, birds and reptiles enclose their developing embryos in an eggshell.

By cell division, animals become a complex system consisting of billions or trillions of specialized cells. Animals will continue to grow and develop, and some may be very dependent upon the parents for their survival.

However, animals all undergo similar patterns of change. We can recognize what kind of animals they are even when they're very young.

A fluid-filled sac surrounds the embryo. It is connected to the mother by a placenta (plə sen´ tə), which has many blood vessels. The placenta is the lifeline for the developing embryo. It carries food and oxygen from the mother's blood to the embryo and empties the embryo's wastes into the mother's blood.

Placenta

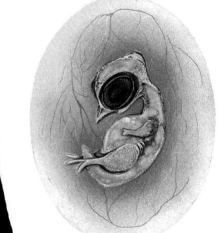

The shell holds in the life-giving fluids but is porous enough to allow gases to diffuse in and out. It also contains food in the form of a yolk and a clear protein you call the "egg white." Blood vessels within the egg carry oxygen, food, and wastes, just as they do in mammal embryos.

GROWING, GROWING, GROWN

Puberty occurs at different times in a female and male.

For people, the stages of development to adulthood continue over many years. And, though they always occur in a regular order, the stages can occur at different times for different individuals. You're familiar with this variety and can see it easily among your classmates. Have you ever heard someone ask, "Why is he or she taller than I am already? I'm older!"

One developmental stage that humans go through as they reach their teenage years is puberty. **Puberty** (pū´ bər tē) is the stage during which your body matures physically. It is marked by changes in physical appearance and is the stage during which your body becomes able to reproduce.

Other stages of development occur throughout a person's life. For example, a woman will ultimately reach an age at which she becomes unable to reproduce. The cycle that caused her body to release an egg each month will stop. This is one more stage in development. Each stage is set at birth by codes in the genetic material from the parents, but each stage may occur with somewhat different timing for different individuals.

Hereditary material within your cells causes you to go through each developmental stage of your life. However, your environment and nutrition also play a role in forming you. The food you eat must supply the cells with proteins, carbohydrates, fats, minerals, and vitamins. So eat healthfully and get plenty of exercise. You can help see to it that your body develops according to the plan your genes have!

A circadian rhythm affects your sleep-wake cycle; most people average about 8 hours of sleep and 16 hours of waking.

Evidence supports the idea that plants and animals both have built-in biological clocks that regulate many of their activities, such as growth, reproduction, and cycles of sleeping and waking. These biological clocks are "set" by genetic codes, and they help living things react to their environments in the most effective ways.

Most of the activities that are affected occur in a 24-hour cycle. Their rhythmic repetition around the planetary day is called circadian (sûr kā´dē ən) rhythm. *Circadian* is Latin for "about a day." For example, certain flowers open and close at the same time each day. Your body temperature exhibits a rhythm—it is probably highest late in the afternoon and lowest just before dawn. Try plotting your circadian pattern for speed, accuracy, and attention span in the following activity.

TRY THIS

Activity!

How Is Your Biological Clock Set?

What You Need

deck of cards, watch or clock with a second hand, *Activity Log* page 33

Shuffle the cards well. Note your starting time, and sort the deck into the 4 suits. Immediately note your finish time. Calculate how long it took and record the time. Repeat this activity 4 times during the day—early in the morning, at midday, in the afternoon, and right before going to bed. Record your results in your ***Activity Log***.

Compare your results to those of your classmates. What conclusions can you make from your data? What is the best time for you to study or take tests? Explain.

REPRODUCING IN STAGES

Some animals, like the butterflies you observed in the Explore Activity, go through the stages of metamorphosis as they grow and develop into adults. No two stages look alike, and all are quite different from the adult form of the animal.

A frog, for example, also goes through different stages in its development. It starts life as a fertilized egg. The egg hatches and out comes a tadpole. The tadpole can swim to small plants on which it feeds. The tadpole continues growing, and its body changes. Legs appear, lungs begin to develop, and teeth grow in the tadpole's mouth. It is beginning the change that will enable it to live on land.

What happens to the tadpole's tail? Since the animal has grown legs, it won't need its tail any more for movement. Does the tail just fall off? No. It becomes smaller and smaller as it is gradually absorbed back into the young frog's body.

After several months, the body of the young frog has metamorphosed completely. The animal has reached its adult form. The pattern of change is complete, and the cycle can begin again when the adult reproduces.

Insects undergo metamorphosis in slightly different forms. The butterfly you saw emerge from its chrysalis had undergone complete metamorphosis, passing through four stages—egg, larva, pupa (pū′ pə), and adult.

The fertilized eggs begin to develop in a jelly-like mass anchored in the water. The eggs hatch in a week or so, and out come tadpoles.

The tadpole develops a long tail and gills so that it can move, take in oxygen, and get rid of wastes in the water.

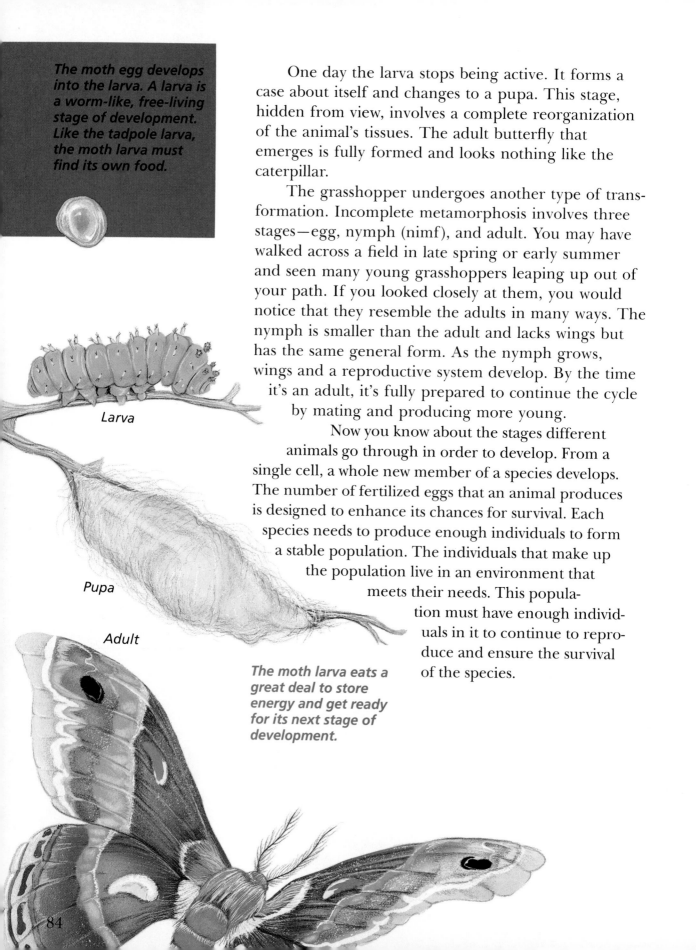

One day the larva stops being active. It forms a case about itself and changes to a pupa. This stage, hidden from view, involves a complete reorganization of the animal's tissues. The adult butterfly that emerges is fully formed and looks nothing like the caterpillar.

The grasshopper undergoes another type of transformation. Incomplete metamorphosis involves three stages—egg, nymph (nimf), and adult. You may have walked across a field in late spring or early summer and seen many young grasshoppers leaping up out of your path. If you looked closely at them, you would notice that they resemble the adults in many ways. The nymph is smaller than the adult and lacks wings but has the same general form. As the nymph grows, wings and a reproductive system develop. By the time it's an adult, it's fully prepared to continue the cycle by mating and producing more young.

Now you know about the stages different animals go through in order to develop. From a single cell, a whole new member of a species develops. The number of fertilized eggs that an animal produces is designed to enhance its chances for survival. Each species needs to produce enough individuals to form a stable population. The individuals that make up the population live in an environment that meets their needs. This population must have enough individuals in it to continue to reproduce and ensure the survival of the species.

The moth egg develops into the larva. A larva is a worm-like, free-living stage of development. Like the tadpole larva, the moth larva must find its own food.

Larva

Pupa

Adult

The moth larva eats a great deal to store energy and get ready for its next stage of development.

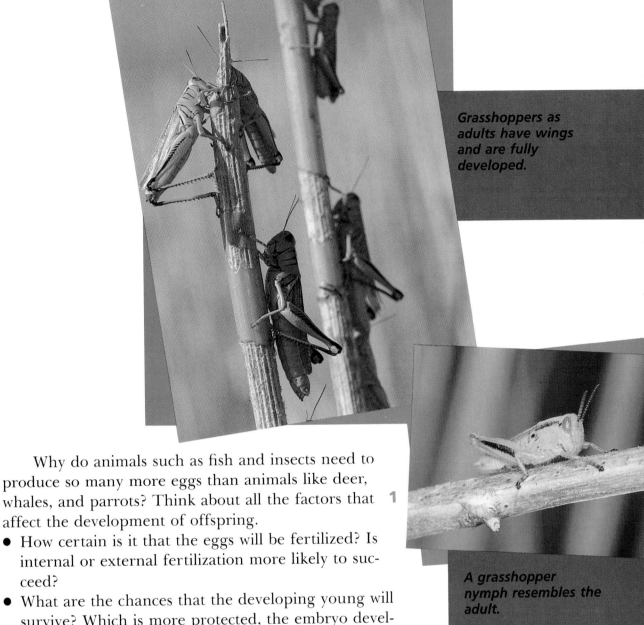

Grasshoppers as adults have wings and are fully developed.

A grasshopper nymph resembles the adult.

Why do animals such as fish and insects need to produce so many more eggs than animals like deer, whales, and parrots? Think about all the factors that affect the development of offspring.

- How certain is it that the eggs will be fertilized? Is internal or external fertilization more likely to succeed?
- What are the chances that the developing young will survive? Which is more protected, the embryo developing in its mother's body or the one developing in water or sand?
- How many of the young will become food for other animals?
- How many times in their lives do the adult animals produce young?

All of these factors influence the survival of a species. Through time, animals have developed methods of reproduction that enable the survival of their species.

CHANGING ANIMAL
POPULATIONS

Minds On! Various stages of an animal's development may mean differences in its structure and its effect on the environment in which it lives. Make a list of animals in your *Activity Log* on page 34. Note several on your list that have an important effect on human populations. At what stage of development do they have this effect? For example, adult female cows produce milk when they calve, and people put this process to work for themselves. Write a paragraph in your *Activity Log* for each animal, explaining how its reproduction, growth and development, or population size can help or harm people. ●

By raising butterflies in the wild, humans are increasing the population and at the same time providing much-needed income for ranchers.

The larvae of the gypsy moth have destroyed millions of acres of forest in the Eastern United States.

HOME, HOME ON THE BUTTERFLY RANCH

SCIENCE TECHNOLOGY AND Society

What happens when the environment of a population is changed or when most of the members of a population die? People are discovering the answer as they continue to take more and more land for agriculture, industry, and residences. When animals lose their habitat or when their habitat becomes too polluted, fewer animals survive, and those that do have a harder time finding each other to reproduce. In some cases, pollution affects their body systems and makes their offspring weak or deformed. The few remaining adults grow older and less fit to reproduce. In extreme cases, the populations become too small to survive. The species becomes extinct.

More and more species' populations are becoming dangerously low. But in some regions of the world, people are coming up with ways to help threatened animals increase their populations. In the Irian Java province of Indonesia, the birdwing butterfly may soon be in jeopardy because habitat areas are beginning to be cleared for agricultural uses. If people clear the land used by the butterflies for habitat and also "harvest" and sell large numbers of the butterflies, the species will soon face extinction.

To avoid this possibility, wildlife experts are working cooperatively with local leaders. A pilot program of butterfly ranching has begun. Along with the ranch land, certain areas are marked off as reserves. In these areas, there is to be no building or farming, and hunting of certain species is restricted.

This program balances the need to protect a species with the need of local people to make a living. This carefully planned use of natural resources is a promising compromise. It seems a realistic way to maintain species' populations in a world where habitats are continually shrinking.

By studying the asexual reproduction of simple animals, scientists have learned how to produce clones of a more complex animal—the frog. A frog's egg cell is fertilized artificially by a specially-treated nucleus from another cell in that frog's body. This cell is removed from the frog's intestine and zapped with a laser. After the egg cell and the altered nucleus unite, a frog embryo develops normally.

The ability to produce clones, or genetically identical organisms, by artificial means leads to some obvious and some not-so-obvious questions. First, can other animals be cloned? Imagine being able to produce an exact copy of a prize steer! Second, should animals be cloned in large numbers? What advantages and disadvantages can you foresee? What problems could develop?

Technological advances have also made it possible to engineer sexual reproduction. Scientists are able to remove eggs from female mammals and fertilize them in test tubes. As the embryos begin to form, they are removed from the test tubes and implanted into other females called surrogates. There they develop normally and later are born.

Why are we interested in reproducing one female's offspring in a surrogate's body? In the case of domestic animals, this can be an important breeding tool. For example, a superior milk-producing cow or a rare and valuable racehorse can contribute eggs even though she is too old to bear young. The offspring borne by the surrogate have the desirable genetic traits of the real mother.

THE BETTER ANIMALS CAN ADAPT TO THEIR CHANGING ENVIRONMENT, THE BETTER THE CHANCE FOR SURVIVAL.

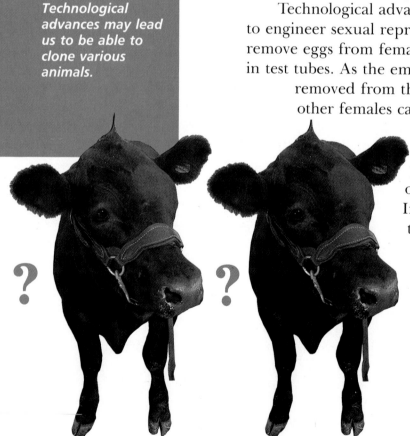

Technological advances may lead us to be able to clone various animals.

Adult animals care for their young to help them survive.

SUM IT UP

The continuation of life is the goal of all living things. When you observed the butterfly chrysalis, you were able to watch one animal in its process of development. This ability of animals to grow and develop is linked to the theory of evolution by the ever-changing patterns and methods that organisms have adopted. The better animals can adapt to their changing environment, the better the chance for their survival. When we study other organisms and their methods of reproduction, we can appreciate any organism no matter how small it may be. Observing how new life forms, develops, and grows, will lead us to new technological advances and preventive measures to fight disease and prolong life.

CRITICAL THINKING

1. What is the difference between a zygote and an embryo?

2. Why is producing a large number of eggs not as important to an animal in which fertilization is internal as it is to an animal in which fertilization is external?

3. When would asexual reproduction be an advantage to a free-swimming flatworm?

4. Many invertebrates are able to regenerate a lost body part. Suggest a reason why all animals are not able to regenerate body parts.

5. In general, how does a newborn human compare with other newborn or newly hatched animals?

ANIMAL ROLES AND

ENVIRONMENTS

How do environments affect the behavior of animals? Why do animals act the way they do?

All animals respond to their environments in ways that are unique to their species. Why does each species have these unique responses? Animals have evolved behaviors that help ensure their survival.

Minds On! What animals can you think of that have developed certain responses to their environments? Make a list of these animals and their responses or adaptations to their environment. Write your ideas in your *Activity Log* on page 35. ●

Animals respond to their environments in an attempt to ensure their survival.

You may have observed an animal's behavior. Have you ever tried to get an animal to change its behavior? When you train a dog to heel or "shake hands," you change its behavior. When you train a cat to scratch its scratching post instead of the furniture, you change its behavior. How do people train animals to behave in certain ways?

Animals respond to their environments only to the level that their bodies will allow. Simple animals have less ability to respond. The more complex the animal, the more developed its body systems are. Therefore, it is able to respond to the environment in different ways. In the following Explore Activity, you'll investigate the responses of fruit flies.

Training animals involves teaching them to act in certain predictable ways.

Activity!

Temperature and Light: How Do Fruit Flies Respond?

Fruit flies are small insects with mouth parts designed for drinking the juices of fruits. Although fruit flies are tiny, they have sense organs that are sensitive to changes in the environment. In this activity, you'll place fruit flies in different environments to see how they respond.

What You Need

2 fruit fly containers with several in each
2 foam vial plugs
container of ice or a refrigerator
black construction paper
masking tape
***Activity Log* page 36**

What To Do

1 Label the containers of fruit flies A and B.

2 Place container A into a container of ice for 10 to 15 minutes. (A refrigerator may be used instead.) Observe the animals' behavior. How has it changed? Write your observations in your ***Activity Log.***

3 Cover one-half of container B with black construction paper.

4 Place both containers in an area where they won't be disturbed until the next day.

5 Count the number of fruit flies in container B that aren't covered by the black construction paper. Record the number in your *Activity Log.*

6 Observe the fruit flies in container A. How has their behavior changed now that they are exposed to room temperature?

7 Return the fruit flies to the place your teacher has chosen.

What Happened?

1. What evidence do you have that fruit flies respond to temperature and light?
2. How do the behaviors of fruit flies help them to survive?

What Now?

1. How do your own responses to temperature and light compare to those of the fruit flies?
2. What animals can you think of that have different responses to temperature and light?

EXPLORE

RESPONSES
ANIMAL

The color of the snowshoe hare depends upon the season of the year.

As you saw in the Explore Activity, when fruit flies were exposed to lower temperatures, their body processes slowed down. They responded to light by moving toward it. Fruit flies' sense organs and other body parts allow them to respond to temperature and light.

Animals may respond to changes in the environment by changing their behavior. Since the environment is always changing, animals must continually adjust or adapt. Some animals may undergo natural body changes. For instance, the snowshoe hare has brown fur during the spring, summer, and early fall and white fur during the winter.

The stability of an animal population is also important. You discovered in Lesson 4 that animal populations remain stable through reproduction. A population is in equilibrium with its environment when it's large enough to reproduce successfully but small enough for all members to share the resources available.

Behavior is an adaptation through which animals can maintain their stability. **Behavior** is the way an organism acts toward its environment. Anything in the environment that causes an organism to react is a **stimulus**. The action of an organism as the result of a stimulus is a **response.**

Many different stimuli affect animals. Some stimuli are outside the animal's body, such as light and noise. Others come from within, such as thirst and hunger. All of these stimuli result in some response by the animal.

Minds On! How many different animal behaviors can you think of? Working in groups of three, brainstorm a list of several animals and the behaviors of each. Next to each behavior, list what stimulus causes the behavior to occur. Write the information in your *Activity Log* on page 38. Share your lists with other groups and discuss the ways you think each behavior helps the animal. Keep this list to use later. ●

All animals respond to stimuli in their environments. Many behaviors of a given type of animal are predictable. Pick up a cat and pet it. It is likely to purr and may knead its paws in your lap. A stranger's car pulls into the driveway. The dog that is usually friendly and playful now barks fiercely. These behaviors are typical of all cats and dogs. Why might that be?

The animal behaviors that usually interest us the most are those of pets and domestic animals. For thousands of years, people have lived closely with certain animals and benefited from them. Why is it useful for us to understand their behaviors?

Knowledge of animal behavior has proved helpful to people by permitting better control of domestic animals such as horses, cows, and goats. When we understand why and when a cow produces the most milk, we can learn how to stimulate greater milk production. Learning more about the behavior of some farm animals can help farmers produce healthy animals that provide better products in larger amounts.

WHEN WE UNDERSTAND WHY AND WHEN A COW PRODUCES THE MOST MILK, WE CAN LEARN HOW TO STIMULATE GREATER MILK PRODUCTION.

Anoles respond by changing colors and blending with their surroundings to avoid predators.

FIELD BIOLOGIST: BIRUTE GALDIKAS

Some scientists are interested in observing animal behavior and understanding it. One such scientist is Birute Galdikas, who has spent the last 20 years studying orangutans in the rain forests of Borneo. Her study has provided us with a detailed description of the orangutan's way of life.

When Birute Galdikas decided to study orangutans, she was working on a master's degree in primatology, the study of primates (mammals that include the great apes, monkeys, and humans). She later earned a doctorate in anthropology.

She spends much of each year following adult orangutans and noting their behavior. They live some 30 meters (99 feet) up in the trees and travel from 500 meters (1,650 feet) to two kilometers (1.2 miles) each day foraging for food.

Her routine begins before dawn and continues the whole day. She notes in her log what the orangutan is doing at one-minute intervals, as well as the exact location of the orangutan in the forest.

Galdikas is also in charge of doing research with some assistants. In addition to the observations of wild orangutans, both Galdikas and her assistants care for

Birute Galdikas studies primates as a field biologist.

a population of about 30 orangutans that have been rescued from illegal captivity. These animals are treated and reintroduced to the wild.

For four months of each year, Galdikas lectures. Through her work she hopes to understand how orangutans live in the wild. She observes how they adapt to their environment, how many young they have, how often they give birth, and how long they live. This information may help us learn more about our own species.

From research like Galdikas's, we have come to understand that animals' actions are a response to something that has happened in their environment. Animal behavior is set up within each animal's genetic codes. From the simplest to the most complex animal, many responses depend upon an animal's genes. Genetic messages limit the body systems and therefore limit the responses of an animal.

ANIMAL BEHAVIORS

Animals have different kinds of behaviors—some they've learned and some they are born with. Do you remember the fruit flies you worked with in the Explore Activity? What caused them to fly down the tube to the light? What caused them to slow down their movement in the cold? These reponses were determined by their genetic makeup. In other words, the behaviors were inborn. A behavior that an animal is born with and that is passed on from generation to generation is an **innate behavior.**

One kind of innate behavior is a reflex. A **reflex** is an automatic response to a stimulus that doesn't involve the brain. Reflex reactions occur because of a fixed pathway along nerve cells that always results in the same response. Shivering, yawning, jerking the hand away from a hot object, and blinking the eyes are all reflex actions. These reactions occur so rapidly that you don't have time to think about doing them before they occur.

The cobra reacts with a reflex to the attack by the mongoose.

Some innate behaviors that help an animal survive are much more involved than reflex actions. A **fixed action pattern** is a complicated series of movements. The acts of the red squirrel seem so complicated and appropriate that they appear to have been learned. Yet, the same kind of squirrels, even if raised in a cage

97

A red squirrel will bury extra nuts in the fall, creating a winter food supply for itself. It takes the nut to a marked place, such as the base of a tree, scratches out a hole, shoves the nut into the hole, and pats dirt over it with its paws.

on a liquid diet, will take the first nuts they are given, carry them to a corner, and make burying motions with their paws. This set of re-actions that help them sur-vive is built into the ge-netic code. It's one of many ways the animals adapt to changes in the environment.

Activity!

Going to Great Lengths

How many kilometers do Arctic terns travel during their migrations?

What You Need
a globe
60 cm of string
masking tape
scissors, ruler
Activity Log page 39

Many young terns hatch in Greenland in late June and early July. Find Greenland on the globe and attach one end of the string there with a piece of tape. When the young terns are ready to begin migration, they leave Greenland and fly to the west coast of Europe. Take the string to France or Spain and tape it again. From there, they fly down the west coast of Africa. Let the string follow this route, taping it where neces-sary. From the southwest coast of Africa, the terns fly over the south Atlantic and the Antarc-tic Ocean to Antarctica. Take the string over this route and tape it. Here is where the birds will stay during the Antarctic summer months. Mark the string where the journey ends.

By May, winter is beginning in Antarctica, so the terns return to Greenland over approximately the same route. Remove the string from the globe and measure a second length equal to the first half of the trip. This represents how far a single tern travels each year. Measure the length of string you used and convert it to km using the scale on the globe. Record your calculations in your *Activity Log*.

Some innate behaviors are very complex, taking the animal great distances to keep it alive. Migration is one such behavior and is thought to be largely innate. The Arctic tern is one bird that migrates. Try calculating just what great distances Arctic terns cover each year in the Try This Activity.

Another kind of behavior that animals have is **learned behavior,** or a behavior that is taught. Learning involves choosing responses to a stimulus, and therefore learned behavior can change as a result of experience and training.

Learned behaviors are not innate but are controlled indirectly by the genetic code each animal has evolved. Genetic material in an animal's cells controls how complex its body will be. Complex body systems—especially a complex nervous system—mean a greater capacity to learn.

Does this mean that only complex animals are capable of learning? Experiments have shown that simple organisms can learn too, at simple levels. Both complex and simple organisms learn through a process of **trial and error,** in which an animal repeats a task, and many times learns from its mistakes. An earthworm placed in a T-shaped maze can learn, after several trials, to turn towards the side with a moist, dark chamber and away from the chamber that is dry and well-lit. After practicing many times, you may have learned to ride a bicycle using the same process.

These Canada geese migrated due to innate behaviors.

Animals can also learn by conditioning. **Conditioning** involves learning a new response in order to receive a reward or avoid a punishment. Dogs are usually taught to do tricks by conditioning. You reward the dog when it does the trick you wish. For example, when it sits, you may give it food, praise it, or pet it. It learns to repeat this behavior in order to receive the reward it wants.

The most complex kind of learned behavior involves insight. **Insight** is the ability to reason and plan a response to a situation. Insight enables some animals to use past experiences to solve new problems. Unlike trial-and-error learning, insight involves the formation of an idea and a plan for testing it. For example, how do experience, an idea, and a plan play a part in the chimpanzee's behavior as pictured here?

In more complex animals, insight is an important source of behavior. Humans are among the most complex animals. Early in their lives, they often learn by trial and error. As they mature, people learn more frequently by insight. Why do you think this is so?

To us, insight means that we can learn not only by our own past experiences but also by the past experiences of others.

No other animals appear as capable of learning as humans. How do you think this has enabled people to dominate Earth?

Conditioned learning involves receiving rewards for jobs well-done.

Minds On! Return to your list of animal behaviors and the stimuli that cause them on page 38 in your *Activity Log*. Decide whether each one involves innate or learned behavior. Label them, and be ready to explain your choices. ●

Science strives to gain knowledge about the world, but it also seeks ways to use that knowledge to improve life. People have not only discovered how different animals learn, they have also found ingenious ways to put this knowledge to work. The U.S. Coast Guard at one time, used reward-punishment conditioning to train pigeons to locate survivors in sea search missions. The pigeons learned that when they saw orange to peck the button in front of them. They would then receive food.

Three trained pigeons were strapped into a strong, transparent, plastic container under a helicopter. Each faced a different way, and each had its own button. When one spotted an orange object in the water, it pecked a button alerting the pilot. Pigeons had very keen vision, and made fewer errors than the best-trained human observers.

Animals behave the way they do for several reasons. Some behaviors are innate, or "programmed" at birth. Other behaviors are learned by trial and error, conditioning, or insight. Many actions are the result of both innate and learned responses. To examine animals in their environments more closely, do the Try This Activity on the following page.

This chimpanzee has learned how to use the stick as a tool for obtaining food.

Activity!

Why Do They Do That?

You can study animal behavior using photographs.

What You Need
magazines or books containing animal photographs, scissors, glue, construction paper or poster-board, *Activity Log* page 40

In magazines or books, find pictures that illustrate different animal behaviors. Clip or copy them. Study each picture. Name the behavior of each animal, and tell whether it is innate, learned, or a combination of both. Mount and label several of your pictures, adding an explanation about the behavior. Describe your 3 favorite photographs in your *Activity Log,* and add your explanations. Compare your responses with those of your classmates. Defend your answers.

Behaviors may also be social. Some animals, such as bees, penguins, humans, and wolves, live and work together in societies. A **society** is a group of animals of the same species living and working together. The animals that live in a society interact in complex ways through social behavior. For example, they establish levels of authority so that there are fewer conflicts. One member of the society is dominant. Other members have less authority, but each member fits somewhere in a chain of relationships. Why might such social structures have evolved in certain animals? How might they help the group survive?

Animals gain many advantages by developing social behaviors. For example, wolves hunt large animals. A pack of wolves is better able to capture a moose than a single wolf is. Musk oxen form herds. With many animals banding together and forming into a circle, it's hard for predators to attack an individual musk ox.

Animals use a number of types of behavior to communicate. **Communication** involves an action by one animal that informs and influences another animal. When we talk about communicating with someone, we're usually thinking about using language. Language is one form of communication, but there are

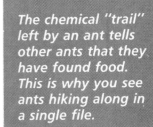

The chemical "trail" left by an ant tells other ants that they have found food. This is why you see ants hiking along in a single file.

many other forms. For example, animals with well-developed eyes may send each other visual messages. Some female butterflies attract male butterflies by the way they fly. Fireflies are attracted to each other by the patterns of their flashes. The intervals of time between flashes is thought to send a specific message. Visual communication is quiet. How does being silent help the communicator survive?

Some animals communicate by means of chemicals called **pheromones** (fâr´ ə mōnz), which have a distinct odor. Some species of insects produce pheromones so powerful that they can communicate over a distance of a kilometer! Pheromones released by some female mammals tell males that the female is ready to mate. Barnacles secrete a pheromone when they attach to a surface. It communicates to other barnacles, inviting them to attach to the same surface. Like other means of communication, pheromones have great survival value. They take very little energy to produce and are not detected by other species. They fit into the plan for survival by allowing animals to respond and reproduce.

103

Animals may also communicate by using sounds. The cricket rubs its front wings against each other to make a chirping sound that sends a message. What other animals can you think of that communicate using sounds?

Many animals have specific behaviors to communicate their claim to a territory. A **territory** is an area that an animal defends from other members of the same species. A male bird may sing or display his bright plumage to declare to other males that this area is his. A clan of howler monkeys howls at other clans that intrude on its territory.

Why do the different members of a population compete for territories? They all have the same needs, and all need a certain amount of space in which to meet those needs. The best-adapted members win territories and attract mates in order to continue their species.

This Lorikeet can defend its territory by singing or showing his colorful feathers.

A wolf leaves scent markers to declare his territory.

Animal behavior helps the animal survive. Why then do birds and mammals play? For example, horses, deer, and antelope chase each other and run just for fun. Bears, wolves, monkeys, and an assortment of other animals play-fight with each other. Ravens have been observed playing games with sticks in the air. They roll in the air, drop the stick, dive and catch it in midair, then repeat the routine. What might be the purpose of animal "fun and games?"

Play is one way young animals might learn. When two squirrel monkeys play-fight, they take turns winning. Perhaps this helps them learn different roles they'll need as adults. They learn that sometimes they should take charge, and at other times they should obey.

It is thought that some animal play teaches the animals how to get along with others. And perhaps, just like humans, animals also play for fun!

In addition to behaviors, animals have developed physical traits that help them survive. These adaptations may include shapes, patterns, or colors that imitate other, more dangerous animals or that even make them resemble plants.

The Australian walking stick looks so much like the twigs it lives on that neither its enemies nor its prey realize that it is there.

105

How have these insects adapted for their survival?

Bumblebee

For example, the chameleon moves slowly but can change quite quickly from bright green to deep brown. This kind of camouflage (kam´ ə fläzh´) is helpful to the insect hunter. The robber fly looks so much like the bumblebee that predators leave it alone. What other animals can you name that mimic something else in order to survive?

Other physical traits that may be less obvious are survival tactics. For example, the elephant's large size protects it. Why? Very few predators will attack such a huge animal. Speed is a great asset to the pronghorn antelope, which can escape an enemy at the rate of 70 kilometers (about 42 miles) per hour. What physical traits have the armadillo and the skunk evolved that help them survive?

Robber Fly

Great Blue Heron

THE PERFECT ANIMAL

When you study animals closely, most reveal a number of unique adaptations to their environments. For example, a heron has long, slender legs and a long beak adapted for wading and spearing prey in the water. The anteater has a long snout and sticky tongue adapted for rooting out and eating ants. The electric eel of the Amazon generates electricity in special muscles. It can jolt an enemy with up to 650 volts!

Choose an animal that you think is perfectly adapted to its environment. List at least five reasons why you think so. Write a persuasive paragraph in which you make your case for your animal.

The government of Kenya burned this ivory to demonstrate its determination to protect elephants from poaching.

ANIMALS

All animals continually change to adjust to changes in the environment that surrounds them. People use this knowledge to help animal populations become more stable. They can even assist them in the fight for survival.

GLOBAL PERSPECTIVE

A FIGHT FOR LIFE

In Kenya, the elephant population has dwindled to a dangerously low level because of illegal hunting by poachers. Despite the fact that the species is endangered, some people kill the animals for their tusks. Ivory tusks are in great demand for jewelry, art, and other uses.

As a result, the African elephant population fell from 1,500,000 in 1979 to fewer than 625,000 in 1989. Many bulls were killed for their large tusks, so fewer females could mate. This caused the population to dwindle even more. With fewer adults, young elephants lacked protectors who could teach them survival skills. The long-range outlook for the African elephant looked bleak.

But people all over the world learned of the animals' plight and responded. The United States, Japan, and a dozen other nations declared bans on the import of ivory in 1989. In the few months immediately following the bans, comparatively few elephants were killed. In a dramatic gesture, the President of Kenya burned tons of tusks that had been taken from poachers. He was demonstrating that the price of ivory was too high if it meant the extinction of the African elephant.

Many elephants are also given protection in Kenya's wildlife preserve, an area in which people may observe animals but may not hunt them. It is hoped that these measures will save the elephants.

One person who assists animals in their fight for survival is a Game Warden. A Game Warden can regulate the way people affect the populations of various types of animals.

Ivory is obtained from the tusks of elephants. Is it worth the price?

The job of a game warden is to protect wildlife by enforcing laws and helping the public to understand and obey wildlife laws. An important task is to see that endangered animals are not hunted illegally. Because poaching (capturing or killing animals illegally) of certain species has become a profitable business, sometimes carried on by crime rings, this job has become more complicated in recent years.

A game warden's job begins with knowledge about environmental science and animal behavior. She or he must know all about the animals and the terrain of their territory. It also helps to enjoy being outdoors. A game warden finds the work satisfying because it supplies a connection with the environment and at the same time protects endangered animals. But beyond that, a game warden needs expertise in crime lab techniques and detective skills. Since the job is much like a police officer's, he or she needs to be disciplined and calm in stressful situations.

Some of the equipment a modern game warden uses to track down and catch poachers in the act includes night-vision goggles, strobe lights with infrared rays that can't be seen by the unaided eye, and video cameras with supersensitive film. With this film, a photo can be taken from more than a mile away, giving proof of illegal activity.

A game warden might also go up in a helicopter or a slow-speed aircraft capable of pinpointing the exact location of a poaching operation by night. If a spotlight (or even a flashlight beam) is seen on the ground, the game warden then radios the position of the beam to agents on the ground. They then apprehend the poacher.

Some game wardens are hired especially to work as undercover agents. Where poaching and selling animals for profit is suspected, the undercover agent poses as a buyer of the protected species and seeks evidence against the criminals.

Protecting endangered species takes some game wardens even further—into the forensics (fə ren´ siks) laboratory. A forensics lab studies evidence to help the government prosecute poachers. In a wildlife crime lab, wildlife parts or products made from animals are traced back to identify the species. Also, the cause of an animal's death may be determined, and a suspect may be connected to the criminal act of poaching.

As people have become more aware of the importance of saving animals from extinction, they have devised various ways of protecting fragile animal populations. At times, this protection can be at odds with land development. For example, a dam may be planned and work begun. However, if an environmental-impact study shows that construction of the dam will destroy the habitat of an endangered species, work may be halted.

It is necessary to work out a plan that allows for the animals' needs as well as the needs of people. This will provide for the stability of all animal life on Earth.

Black rhinos found in Africa have been poached and their numbers have been greatly reduced.

Sum It Up

The adaptations of animals have allowed them to survive and become better adjusted to their environments. By observing how animals respond and adapt, as with the fruit flies, you can learn from them and learn more about how animals evolve. The physical traits of animals allow them to carry out certain behaviors and live in certain environments. Animals have continued to evolve, and their adaptations have made them better able to survive in today's world. All animals (including humans) will continue to thrive if care is taken not to destroy the environments to which animals have been adapting for millions of years.

Critical Thinking

1. Why would you expect the behavior of a chimpanzee to be more complex than that of an earthworm?
2. What advantages might a wolf have by living in a pack of wolves rather than by living alone?
3. How are behaviors valuable adaptations just like shapes of beaks or feet?
4. How do you think a society organized by dominance helps a species survive and reproduce?
5. In what ways does communication help an animal species survive?

ANIMALS AND SCIENCE

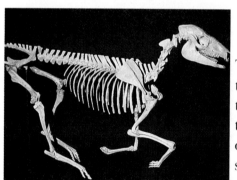

You've learned a great deal about animals. Through your studies, you've felt the struggle of living things to survive. You've observed the different ways that animals reproduce. Reproduction ensures that there will always be more of an animal's kind. You've observed how animals survive by the ways they respond to the environment around them.

Scientists have observed various animals closely. They have studied the history of Earth as it's written in layers of rock and the history of animals as it's preserved in fossils, traces of life from the past. What they have learned from this evidence has led to a theory of how life began and how it evolved into many different forms.

Since Earth itself is always changing (just as it has been for billions of years), living things must always change to stay alive. The theory of evolution helps to explain the diversity and overwhelming abundance of life on Earth.

So far, much evidence supports the mechanisms of evolution. Using scientific methods, scientists continue to try to understand how evolution leads to the diversity of life. The debate over evolution is an ongoing process, full of active exploration, changing ideas, and opportunities for discovery. This is the nature of science, the best tool we have for studying and understanding the lives of animals.

The goal is to add to our store of knowledge about the world of animals. Some of our knowledge is put to use to improve our world. Perhaps most importantly of all, we have learned that humans are only one species among many, all of whom struggle to survive, and all of whom are important in the network of interdependent living things.

Minds On! You've learned a great deal about the world's animals. Put to work what you have learned by using your imagination. Imagine an animal no one has discovered yet. Perhaps it lives in a hidden region on Earth. Perhaps it exists on another planet or in another dimension.

You discover it. Imagine your experience. What is the animal like? How intelligent is it? How does it react to you?

Pretend you are a scientist observing a group of these animals. Report on the society or civilization the animals have. Describe the animal considering these points:

- what it looks like
- what body systems it has and how they work
- its adaptations to its environment
- how its young are produced, grow, and develop
- its chances for survival

Decide on how you'll show your imaginary animal to your classmates. Will you sculpt the animal out of clay, write a report, give a speech, draw pictures? If you wish, work in a small group to present classmates' animals and your own. ●

Animals are always changing and adapting to their environments. The theory of evolution helps to explain why.

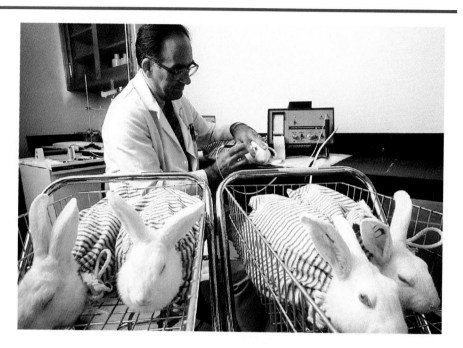

For hundreds of years, human civilization has depended heavily upon scientific progress. Research is important, and some animals are commonly used in scientific research. Yet many people, known as animal-rights activists, believe animals have just as much right as people to live pain-free, unconfined lives. These activists are angered by the pain, abuse, and exploitation they feel are involved in animal research.

For example, they point to the use of laboratory animals in tests of chemicals. Rats may be given massive doses of chemicals that are suspected of causing cancer. Often, they die. Experimental medicines and cosmetics are tested on animals first, before being tested on humans and marketed. A new lotion may be tested in a rabbit's eyes to make sure it is safe.

Some testing of animals is simply to gain knowledge without any immediate practical application. A scientist may operate on a cat's brain in order to learn more about how the brain functions. The cat may be permanently affected or may even die.

Most laboratory testing is motivated by concerns for human welfare. Where do the animals' rights come in? Are we willing to slow research or risk unsafe products by doing away with animal testing? Are there alternatives to the kinds of animal testing currently being used? How else might scientists run safety tests and test new materials?

Animal-rights activists also protest the widespread killing of some animals by hunting and fishing. Many young harp seals are killed each year for their fur. Fishermen netting tuna may accidentally kill dolphins who swim into their nets.

On the other hand, some human societies are based entirely on hunting or fishing. The hunters or fishers sell the meat, fish, or furs of the animals they kill. If they cannot hunt or fish, they have no way to make a living. And it is also true that limited hunting or fishing is good for a population of animals or fish. It prevents overpopulation that can lead to crowding, disease, and starvation for animals.

Where do you stand on the issue of animal rights? Research these issues and make plans for a class debate. How will you defend your stand on the issue of animal rights? Can you support your position with facts? Write your defense and be prepared to present it to the class.

GLOSSARY

Use the pronunciation key below to help you decode, or read, the pronunciations.

Pronunciation Key

a	at, bad	d	dear, soda, bad
ā	ape, pain, day, break	f	five, defend, leaf, off, cough, elephant
ä	father, car, heart	g	game, ago, fog, egg
âr	care, pair, bear, their, where	h	hat, ahead
e	end, pet, said, heaven, friend	hw	white, whether, which
ē	equal, me, feet, team, piece, key	j	joke, enjoy, gem, page, edge
i	it, big, English, hymn	k	kite, bakery, seek, tack, cat
ī	ice, fine, lie, my	l	lid, sailor, feel, ball, allow
îr	ear, deer, here, pierce	m	man, family, dream
o	odd, hot, watch	n	not, final, pan, knife
ō	old, oat, toe, low	ng	long, singer, pink
ô	coffee, all, taught, law, fought	p	pail, repair, soap, happy
ôr	order, fork, horse, story, pour	r	ride, parent, wear, more, marry
oi	oil, toy	s	sit, aside, pets, cent, pass
ou	out, now	sh	shoe, washer, fish mission, nation
u	up, mud, love, double	t	tag, pretend, fat, button, dressed
ū	use, mule, cue, feud, few	th	thin, panther, both
ü	rule, true, food	th	this, mother, smooth
u̇	put, wood, should	v	very, favor, wave
ûr	burn, hurry, term, bird, word, courage	w	wet, weather, reward
ə	about, taken, pencil, lemon, circus	y	yes, onion
b	bat, above, job	z	zoo, lazy, jazz, rose, dogs, houses
ch	chin, such, match	zh	vision, treasure, seizure

amphibian (am fib′ ē ən): a cold-blooded vertebrate that has a moist skin with no scales and lives part of its life in water and part on land.

arthropod (är′ thrə pod′): an invertebrate that has jointed legs, an exoskeleton, and a segmented body.

asexual reproduction (ā sek′ shü əl rē prə duk′ shən): the production of offspring from only one parent; *see also* sexual reproduction.

axon (ak′ son): a nerve fiber that carries impulses from the cell body of one neuron to the next.

behavior (bi hāv′ yər): the reaction of an organism to its environment.

bird: a warm-blooded vertebrate with wings, a beak, two legs, and a body covered with feathers.

bone: a hard substance containing calcium and phosphorous that makes up the skeletal system.

book lung (bu̇k lung): a breathing structure of an arachnid that consists of thin leaves of tissue that resemble pages of a book through which air passes.

carnivore (kär′ nə vôr′): an animal that feeds on plant-eating animals and other carnivores.

circulatory system (sûr′ kyə lə tôr′ ē sis′tʃm): transport system made up of the blood, blood vessels, and heart that circulates blood throughout the body.

closed circulatory system transport system in which the blood remains inside blood vessels that circulate it throughout the body; *see also* open circulatory system.

cnidarian (nī dâr′ ēn): a hollow-bodied animal with stinging cells.

cold-blooded: having a body temperature that varies with the temperature changes of the environment.

communication: the exchange of signals or messages such as flashing lights, odors, bright colors, or sounds performed by one animal that informs or influences another animal.

complete metamorphosis (kəm plēt′ met′ ə môr′ fə sis): development in animals that involves a process consisting of four stages: egg, larva, pupa, and adult.

conditioning: training to cause a response to a stimulus that does not normally cause that response.

dendrite (den′ drīt): a branch of a neuron that receives stimuli.

diaphragm (dī′ ə fram′): a sheet of muscle across the bottom of the chest cavity that expands and contracts to move air into and out of the lungs.

digestive system (dī jes′ tiv sis′ tʃm): a group of organs that act together to take in food and change it to a form cells of the body can use.

echinoderm (i kī′ nə dûrm′): a marine invertebrate that has a mineral skeleton with spines.

endoskeleton (en′ dō skel′ i tʃn): a skeleton located inside the body.

excretory system (ek′ skri tôr′ ē): a group of organs that act together to eliminate wastes from the body.

fertilization (fûr′ tə lə zā′ shən): the process in which a male and a female gamete combine to form a zygote.

fish: a cold-blooded vertebrate that lives in water and obtains oxygen from the water by using gills.

fixed action pattern: a complicated series of movements.

flatworm (flat′ wûrm′): a simple animal that has a flat, legless body with an identifiable head and tail.

gestation (jes tā′ shən): the time between fertilization and the birth of an offspring.

gill (gil): an organ found in fish used to obtain oxygen from water and release carbon dioxide from the blood into water.

herbivore (hûr′ bə vôr′): an animal that eats only plants.

incomplete metamorphosis (in′ kəm plēt′ met′ ə môr fə sis): development in animals that involves three stages: egg, nymph, and adult.

innate behavior (i nāt′ bi hāv′ yər): a behavior that an animal is born with, which is passed on from generation to generation.

insight (in′ sīt′): the ability to reason and respond to a situation.

learned behavior (lûrnd bi hāv′ yər): a behavior that is taught through experience or training.

mammal (mam′əl): a warm-blooded vertebrate that has hair and produces milk to feed its young.

mantle (man təl): a fold of tissue in mollusks that makes the shell.

marsupial (mär sü′ pē əl): mammal whose young complete their development in the mother's pouch.

meiosis (mī ō′ sis): a method of cell division in which sex cells are produced.

metamorphosis (met′ ə môr′ fə sis): the transformation from larva to adult form that many invertebrates and amphibians undergo.

mitosis (mī tō sis): the process in which a cell's nucleus divides forming two new cells with identical genetic material.

mollusk (mol′ əsk): soft-bodied invertebrates that live on land or in fresh or salt water.

muscular system (mus′ kyə lər sis′ təm): a group of organs that act together to allow for movement of an animal.

nervous system (nûr′ vəs sis′ təm): a group of organs that act together to regulate and control the activities of an animal.

omnivore (om′ nə vôr′): an animal that feeds on both plant and animal material.

open circulatory system: a system of transport in which blood is not contained in vessels but circulates freely through the body surrounding the cells; *see also* closed circulatory system.

pheromone (fer′ ə mōn′): a chemical that conveys information to other members of a species.

placenta (plə sen′ tə): sac that grows into the wall of the uterus through which food, oxygen, and wastes move back and forth between mother and developing young.

puberty (pū′ bər tē): the stage of development during which the body becomes physically able to reproduce.

radula (raj′ ə lə): a tongue-like organ of some mollusks, which has rows of teeth for tearing and scraping food.

reflex (rē′ fleks′): an automatic response to a stimulus.

regeneration (ri jen′ ə rā′ shən): the regrowth of lost or damaged tissues and organs.

reptile (rep′ təl): a cold-blooded vertebrate that has scales, breathes air, and lives mainly on land.

respiratory system (res′ pər ə tôr′ ē sis təm): a group of organs that act together to bring oxygen into the body and expel carbon dioxide from the body.

response (ri spons′): a change in behavior as a result of a stimulus.

roundworm (round′ wûrm′): an invertebrate that has a round, tube-like body tapering to a point at each end.

segmented worm (seg′ mənt əd wûrm): a worm that has a tube-shaped body divided into segments that are similar in structure.

sexual reproduction (sek′ shü əl rē prə duk′ shən): the production of offspring using sex cells.

skeletal system (skel′ i təl sis′ təm): a group of organs that act together to provide support and protection to the body and allow the body to move.

society (sə sī′ i tē): a group of animals of the same species that interact together.

sponge (spunj): the simplest kind of animal, having no definite shape and living attached to one spot.

stimulus (stim′ yə ləs): something in the environment that causes an animal to react.

synapse (sin′ aps): the space between neurons.

territory: area that an animal or group of animals defends against others of its species.

trial and error: a type of conditioned learning in which an animal develops a behavior by learning to avoid mistakes.

verebra (vûr′ tə brə): one of a series of bones that make up the spine; plural: *vertebrae.*

warm-blooded: maintaining a nearly constant body temperature not influenced by the temperature of the surroundings.

zygote (zī′ gōt): a single cell produced by fertilization that grows by cell division to become a complete organism.

INDEX

INDEX

CREDITS

Cover, ©Guido Alberto Rossi/The Image Bank; 1, ©Gerry Ellis/The Wildlife Collection; 3, (t) ©KS Studios/1991; (b) ©Gerry Ellis/The Wildlife Collection; 4, (b) R. Becker, PhD/Custom Medical Stock; (t) Ken Deitcher/The Wildlife Collection; 5, (l) Richard Hermann/The Wildlife Collection; (r) E.R. Degginger/Color-Pic; 6, (t) ©Ardea; (b) ©Ron and Valerie Taylor/Bruce Coleman, Inc.; 7, (t) ©Denise Tackett, Tom Stack & Associates (b) ©Heather Angel/Biofotos; 11, ©Studiohio, 1991; 12, ©Chris Huss/The Wildlife Collection; 13, (t) ©E.R. Degginger/Color-Pic; (c) ©Larry Jensen/Visuals Unlimited; (b) ©Phil Degginger/Color-Pic; 14, (l) ©E.R. Degginger/Color-Pic; (r) ©M.I. Walker/Photo Researchers, Inc; 16-17, ©Chris Huss/The Wildlife Collection; ©Gary Bell/The Wildlife Collection; 20, ©William J. Weber/Visuals Unlimited; 22, (t) ©Evelyn Tronca/Tom Stack & Associates; (bl) ©William J. Weber/Visuals Unlimited; (br) Studiohio/1991; 23, (tr) ©C.P. Hickman/Visuals Unlimited; (cl) ©Gerry Ellis/The Wildlife Collection; 24, (t) ©Brian Parker/Tom Stack & Associates; (b) ©William E. Ferguson; 26-27, ©Carl Roessler/Animals Animals; 28, (l) ©John Serrao/Visuals Unlimited; (r) Jason Laure; 29, ©Tom McHugh/Allstock; 30, (l) ©Kenneth W. Fink/Bruce Coleman, Inc.; (r) ©Len Rue, Jr./Bruce Coleman, Inc.; 31, (t) ©Martin Harvey/The Wildlife Collection; (c) ©Carl Roessler/Bruce Coleman, Inc.; (b) ©E.J. Maruska/Visuals Unlimited; 33, (t) ©Superstock; (b) ©Elaine Shay; 34, (t) ©Zig Lesczynski/Animals Animals; (b) © Ed Robinson/Tom Stack & Associates; 35, ©Runk/Schoenberger/Grant Heilman Photography; 36, (t) ©Ken Deitcher/The Wildlife Collection; (b) William E. Ferguson; 37, (t) ©Hal Harrison/Grant Heilman; (it) ©Frans Lanting/Allstock; (b) ©John Cancalosi/Tom Stack & Associates; 38, ©Jeff Foott/Tom Stack & Associates; 39, (t) ©David G. Barker/Tom Stack & Associates; (it) ©Zig Lesczynski/Animals Animals; 40, ©Tom McHugh/Allstock; 41, (t) Paulette Brunner/Tom Stack & Associates; (it) ©Grant Heilman Photography; 42, ©Brent Turner/BLT Productions/1991; 43, (t) ©Roger and Donna Aitkenhead/Animals Animals; (b) ©William E. Ferguson; 44, (t) ©E.R. Degginger/Animals Animals; (b) ©H. Reinhard/Allstock; 45, ©Gerry Ellis/The Wildlife Collection; 46, (t) ©Dave Watts/Tom Stack & Associates; (b) ©C. Andrew Henley/Biofotos; 47, (t) ©Art Wolfe/Allstock; (b) ©William E. Ferguson; 48, ©Bauer/Allstock; 50-51, ©R. Becker, PhD./Custom Medical Stock Photo; 51, ©Ron Spomer/Visuals Unlimited; 52-53, ©Studiohio/1991; 54, ©John Cancalosi/Natural Selection; 55, ©L. West/Photo Researchers, Inc.; 57, (t) ©Margot Conte/Animals Animals; (b) ©E.R. Degginger/Color-Pic; 58, (t) ©Denise Tackett/Tom Stack & Associates; (c) ©Tom McHugh/Photo Researchers, Inc.; (b) ©Steve Simonsen/Natural Selection; 61, (tr) ©Ken Deitcher/The Wildlife Collection; (bl) ©Richard Parker/Photo Researchers, Inc.; 62, (tl) ©Brian Parker/Tom Stack & Associates; (b) ©Martin Dohrn/Photo Researchers, Inc.; 65, ©Runk/Schoenberger/Grant Heilman Photography; 66, ©Patricia Barber/Custom Medical Stock Photo; 67, ©Martin Harvey/The Wildlife Collection; 70-71, ©Jerome Wexler/Photo Researchers, Inc.; 70, ©Heather Angel/Biofotos; 71, (tc) ©Martin Harvey/The Wildlife Collection; (bl) ©Nada Pecnik/Visuals Unlimited; 72-73, ©KS Studios/1991; 76, (t) ©Cabisco/Visuals Unlimited; (b) ©Richard Herrmann/Wildlife Collection; 77, (tl) ©Barry L. Runk/Grant Heilman Photography; (tr & br) ©George Anderson/1991, (r) ©Paolo Curto/The Image Bank; 79, (tr) ©Rick Hall/Custom Medical Stock Photo; (br) ©Patricia Barber/Custom Medical Stock Photo; 81, ©KS Studios/1991; 82, (l) ©George C. Anderson/1991; (r) ©Bonnie Rauch/Photo Researchers, Inc.; 85, (l) ©Grant Heilman Photography; (r) ©Bruce Gaylord/Visuals Unlimited; 86-87, ©E.R. Degginger/Animals Animals; 86, ©E.R. Degginger/Color-Pic; 87, ©E.R. Degginger/Animals Animals; 88, ©Jay Brousseau/The Image Bank; 89, ©Will Troyer/Visuals Unlimited; 90-91, ©Anup an Manoj Shah/Animals Animals; 90, ©Gerry Ellis/The Wildlife Collection; 91, ©KS Studios/1991; 92-93, ©Studiohio/1991; 94, (tl) ©Michael Francis/The Wildlife Collection; (bl) ©Robert Laniken/The Wildlife Collection; 95, ©E.R. Degginger/Color-Pic; 96, ©Christopher Springmann; 97, (t) ©Kennan Ward; (b) ©E.R. Degginger/Color-Pic; 98, ©Michael S. Quinton/Visuals Unlimited; 99, ©Gerry Ellis/The Wildlife Collection; 100, ©Superstock; 101, ©Gerry Ellis/The Wildlife Collection; 102, ©KS Studios/1991; 103, (t) ©Gerry Ellis/The Wildlife Collection; (b) ©KS Studios/1991; 104, (t) ©Gerry Ellis/The Wildlife Collection; (b) ©Michael Francis/The Wildlife Collection; 105, ©Gerry Ellis/The Wildlife Collection; 106, (t) ©E.R. Degginger/Color-Pic; (b) ©Gerry Ellis/ The Wildlife Collection; 107, ©Steve Jackson/Black Star; 108, ©Don W. Fawcett/Visuals Unlimited; 109, ©E.R. Degginger/Color-Pic; 110, A. Kerstitch/Bruce Coleman, Inc.; 112, ©Hank Morgan/Photo Researchers, Inc.; 113, ©David Butz/FPG International.

Illustration Credits:

7, 8 (tl), 110, 111, Jean Cassels; 8 (b), Amanda Root; 9, 69, Bill Singleton; 19, 21, 23, 25, 50, 51, 54, 55, 56, 58, 61, 63, 64, 74, 75, 80, 83, 84, Lee Mejia; 32, 35, 40, 59, 60, 78, 79, Felipe Passalacqua; 42, Anne Rhodes; 70, 71, Jan Wills.